AGRICULTURAL LANDUSE IN INDIA

(A Study in Yashoda Basin)

AGRICULTURAL LANDUSE IN INDIA

(A Study in Yashoda Basin)

B.C. Vaidya

MANAK
PUBLICATIONS PVT. LTD

First Published in 1997
by **MANAK PUBLICATIONS (P) LTD.**
G-19, Vijay Chowk, Laxmi Nagar, New Delhi-110 092
Phone: 2453894 Fax: 91-11-2418599
Branch Office: 3-Na-20, Jawahar Nagar, Jaipur-302004

ICSSR Consultant
Prof. Noor Mohammad

ISBN 81-86562-30-3

Author
B.C. Vaidya

Laser Typeset by
Akshar
Ph: 2478932

Printed in India by
Triveni Offset
Ph: 2288175

To My Teachers ....

Prof. B.G. Tamaskar
Prof. S.C. Gupta
Prof. N.G. Jain
Prin. V.H. Ghorpade
Shri. R.R. Pohanekar

For Replenishing My Faith in the Honest Pursuit of Learning

FOREWORD

Agriculture continues to sustain millions of people in India. This is despite the rapid industrialization which is acquiring significant proportions in many regions of the country. The twin processes of industrialization and urbanization are transforming the traditional economic relations between the rural and the urban segments of population.

Several facets of this changing economic scenario may be noted. First, increase in population in the rural areas, which has maintained a consistently high growth rate, seems to have led to strong rural-urban migration waves, sustained and supported by the rising demand for labour in the urban sector. Secondly, the exodus from the rural areas not only contributed to unprecedented urban growth but also expanded the market for the consumer goods produced by the formal industrial sector. Thirdly, these developments provided an impetus to agriculture which adjusted to the new conditions and responded to the new links now being forged between the urban and the rural segments of economy.

As terms of trade in agriculture improved in many regions of the country, with varying degrees of intensity, significant changes took place in the institutional framework of agriculture resulting in the accumulation of land and other agricultural assets in a few hands. Investment in agriculture to improve irrigation infrastructure or acquire agrarian technology facilitated capitalist development in the farming sector. These changes had their impact on the overall performance of agriculture transforming it eventually into a surplus generating economy.

Understandably the impact of these changes remained confined to the selected regions of the country with the result that the gap between the developing and the under-developed regions widened further. It is these underdeveloped regions where agrarian relations continue to be determined by tradition and where a technological break through seems to be a far cry.

Focusing on one such underdeveloped region in Wardha district of Maharashtra this study presents a profile of the agricultural land use at the village-level. The work follows the tradition of the British geographer. L. Dudley Stamp, who insisted on recording faithfully the existing land use of every inch of soil before any intervention could be made towards its improvement. An Indian student of Stamp, Professor Mohammad Shafi, Continued this tradition pleading vigorously for the urgency of this need for land use studies in the context of deteriorating man-land ratio in India. The Aligarh school under the leadership of Professor Mohammad Shafi conducted land use studies at the micro scale covering different parts of the North Indian Plain, particularly the middle and the upper stretches of the Gange Valley. The underlying assumption of this approach is that in order to ensure the rational utilization of our land resources, the existing land use should be recorded carefully. Using cadastral maps available with the land revenue administration in each district, the researcher should produce a comprehensive picture of the truth at the grass-roots land displaying uses of land from primary classification to cropping pattern. The idea is based on the conviction that it is the land which sustains our population and an efficient use of land can help meet the increasing demand for food and agricultural raw materials as the population growth poses a serious challenge to the food security of the teeming millions. The planning agencies at the national and the state levels, did not show much of an interest in this approach nor did they benefit from the plethora of data, generated through such studies at the grass-roots level. In fact, nothing could be done to check the haphazard growth of cities and their consequent encroachment on the good quality agricultural lands. Industries continued to develop on the prime land best suited for agricultural use. Agricultural planning, like all other planning in this country, suffered from administrative ad hocism, leaving little scope for a meaningful interaction between the academics and the planning administrators. Meanwhile, the rural scene changed as the stronger elements in the rural world, the kulaks, drawn from the middle ranking castes, intensified their stranglehold on land, marginalising further the medium and the small farmers. The process of land accumulation added phenomenally to the rank and file of the agricultural labourers who slided down the poverty line. This added to the social tensions which have become the norm of the day.

Dr. Balkrishna Vaidya has done a good job in selecting Wardha, a backward district of Maharashtra already chosen by the Planning Commission under the integrated rural development programme. In doing so he has drawn our attention to the fact that systematic land use studies have not lost their value. In fact, they should be further expanded so as to incorporate within their matrix the on-going social dynamics of the village society, particularly the processes leading to the deprivation of the vulnerable sections, such as the delits, and the tribes who are waging a struggle for survival. Land use studies can not ignore the imperative of social development.

New Delhi

Aijazuddin Ahmad
Professor of Geography
Jawaharlal Nehru University,
New Delhi.

PREFACE

Agriculture is perhaps the most complex phenomenon. This is because of the large number of geo-hydrological, environmental, biological and socio-economic variables influencing the agriculture production processes. As a result, diversities abound, not only in the cropping systems but in the farming systems as well.

Planning is a continuous process. This is true of macro as well as micro level planning. The concept of agricultural production planning has undergone transformation over the years in India. The micro level agricultural planning has progressed from crop planning to watershed level planning. The cropping system approach has given way to farming system approach. Any such watershed level planning has to take into account the authentic anyalysis of the existing scenario in respect of natural resources and potentialities. Only then, it is possible to devise suitable strategies and draw up development plan for augmentation of resources, infrastructure, services etc.

This book presents an analysis of physical as well as socio-economic variables which impinge on the landuse and cropping systems prevalent in Yashoda basin (Wardha District). Cotton is the key crop having direct bearing on the overall landuse, crop sequence and economic well being of the farmers in the basin. This micro level analysis will provide a benchmark for drawing perspective plan for agricultural development in the Yashoda basin. The author of this book

Dr. Balkrishna C. Vaidya deserves compliments for the foresight and painstaking efforts he has taken for bringing out this valuable piece of information for all those involved in the development task.

JAGDISH PATIL
Additional Director Agriculture
(Input & Quality Control)
Maharashtra State
PUNE-411 001

PUNE
20th Feb. 1997

ACKNOWLEDGEMENT

There are many people who helped in successful completion of this project. I am sincerely thankful to Prof. B.G. Tamaskar for valuable help, guidance throughout the course of during my study. Dr. S.C. Gupta, Late Dr. P.R. Karmarkar, Dr. P.T. Malshe, Prof. E.V. Lingiah, Prof. Kumud Pawade, Prof. N.G. Jain, Prof. K.T. Kolte, Prof. P.G. Kowathalkar, Prof. T.V. Gedam who have always been a source of inspiration in this honest persuit, they deserve special thanks.

I owe thanks to Professor and Head Dr. S.B. Sawant for entending required facilities in our geography department, University of Pune. Dr. V.S. Kale deserves for data processing. The faculty members of our department, namely, Dr. Alka Gadgil, Dr. S.R. Jog, Dr. V.S. Datey, Dr. Jaimala Diddee, Prof. V.S. Deosthali, Prof. V.P. Kairkar for their cooperation. Non teaching staff of department owes special thanks. Thanks are also due to Shri G.D. Walvekar and Shri Shashank Waghmare for cartographic assistance and typing respectively. I wish to place on recorded my appreciation to Shri. Adv. Promod Fulmali, Dilip Gomase, Nana Tadas, Baban Jivatode, Kishor Vaidya, Shekhar Bidwaik, Ramdas Chandekar, Vasantrao Kapse, Kasherao Mangekar, Gulabrao Pohane, Hiraman Rade, Vitthal Kumare, Shri Bansode, Prof. Rajguru, Shri Narayan Kadam, Shri M.G. Shinde, Shri Antawal deserve for thanks. I must acknowledge to Dr. P.J. Meshram, Dr. S.V. Khandewale, Shri R.M. Ramteke, and Shri P.S. Mehta, New Delhi for their help and cooperation. I received admiration and appreciation from my friends and thier families among them who owe due regards, namely, Ramdas Chendekar, Shri Nana Selkar (Teacher) Shri Manohar Bodkhe, Baba Hinge, Prof. V.G. Virulkar. I thanks all of them earenstly. I express my deep appreciation to authorities of I.C.S.S.R. for the financial assistance. I must thanks to Shri S.E. Surve for manuscript typing and Shri M.D. Pathak for ever well wishing and great help in proof reading. My wife Chhaya who is a source of constant help in this study. I also appreciate ever encouragement from Shri Inje family.

There are many well wishers who helped and although they are not mentioned here by name, I thanks them sincerely. Finally, Shri Mathura Prasad Yadav of Manak Publications Pvt. Ltd. for undertaking printing and publishing this book with personal touch with maintaing high quality of the book.

Dr. B. Vaidya
Department of Geography
University of pune
PUNE-411 007

PUNE
20th January, 1997

Contents

1

Introduction

GENERAL INTRODUCTION

Agriculture is one of the oldest economic activities of man. Generally, it is understood to mean both, cultivation of food and fibre crops and the raising of livestocks. It has remained an important source of livelihood even today over the years in spite of growing industrialization and urbanization in the world and nearly fifty per cent working population is still engaged in agriculture (Gigg, 1974). In the developing countries, agriculture sector has been a principal source of employment and largest portion of income. However, it provides raw material to industry and much of export items (Davis, 1982). Agriculture in India is a major source of economy. It contributes fifty per cent to national income and gives direct employment to about 68 per cent of total population and nearly ninety per cent population to rural area (India, 1983) and provides nearly thirtyfive per cent of country's export, besides, supplying of wagegoods required the non-agricultureal section in industry. Hence it is not surprising that agriculture in India has been receiving much attention in the sectoral allocation of inputs in the Five Year Plans in India and therefore top priority has been assigned to develop agriculture sector in our country.

The intensive efforts in India have been made during the Five Year Plans to make agriculture self-reliant through increase in area under plough, yield per hectare and total production by providing schemes, subsidy and many concessions to improve the food production and stress was, primarily, given on improved tillage, adoptation of better cultivation and methods, use of fertilizers and improved seeds and a more plentiful and assured supply of water. Despite several attempts through Five Year Plans in India, the process of agricultural growth has not yet been properly channalized due to lack of proper understanding of basic characteristics, its scientific evaluation and landuse.

The study of land and agriculture from geographical point of view gained importance only after fifties in the context of regional planning. Though more than 150 years ago Von Thunon's famous theory of location was concerned with agriculture, the era as new concept in agricultural geography started only with the works of Jonasson (1925) and Baker (1926). Later on, Hartshorne and Dicken (1935), Whittlesey (1936). The interest shown in landuse and agriculture by Weaver (1944), Peter Scot (1957), Johnson (1958), Stamp (1962), Kostrowick (1957), Lee (1963), and Powel (1969), further opened new vistas of agriculture geography. In India, agriculture geography started rather late. Dr. B.N. Mukharjee (1942), Agricultural region of Uttar Pradesh was the first to attempt in this direction within fouty years. After this many studies in this field were completed and notable contribution was made by Dayal (1950), Shafi (1960), Bhatia (1965), Ayyar (1961), J. Singh (1967), Saxena (1968) and Joshi (1972).

In the field of land utilization and agricultural problems a few attempts have been made in Maharashtra by Kumbhar (1978), Shinde (1989), More (1980), Powar (1984), Sargarh (1986), Karmarkar (1981), and Datye (1984). These studies have evaluated landuse patterns and agricultural problems of western Maharashtra.

CHOICE OF THE REGION

The choice and topic under investigation is influenced by many considerations. Firstly, the study region falls within one of typical agricultural tract of dry farming on the Deccan trap region. To step up the agricultural production in Yashoda basin, no indepth investigations and study have been attempted so far. Secondly, the agricultural landuse study of the region remains still untouched totally for planning and development. Hence the existing landuse of the Yashoda basin has been studied in order to uncover agricultural scenario of this region. This study of landuse has been attempted at village level. The agricultural landuse including its other facets have been attempted here covering the district and the region concerned. The aims and objectives of the proposed study yield to present a concrete and realistic picture of the existing agricultural landuse. The appropriate agricultural landuse of the region under study calls for proper planning to exploit the potential of the region, particularly in the contents the recent improvement and modernisation of agricultural sector. Thirdly, Wardha dstrict in which the Yashoda Basin falls has been already selected by the Planning Commission, Government of India, under the programme of Integrated Rural Development for its all round development.

And fourthly, selection of region calls for such an analytical study as it is economically backward and its basic resources reveal considerable potential. The related and relative infra-structure can yield impressive results. Above all, the close familiarity with the agrarian base of the basin has motivated the researcher to undertake this investigation.

AREA UNDER STUDY

The Yashoda Basin of Wardha district in Maharashtra comprises a region of small areal extent in southwest (Exhibit 1.1) having 155 villages of four talukas, 75 villages from Wardha taluka, 55 villages from Dewali taluka, 17 villages from Hinganghat taluka and 8 villages from Arvi taluka in the district. Except the source region and the north hilly part, the entire Yashoda Basin is a plain. The study region of 860.88 sq. kms. encompasses 13.64 per cent of Wardha district and 0.27 per cent area in Maharashtra having 114306 population according to 1981 census. The medium black to deep black soil appear throughout the region except the north part where hilly region occurs is of rugged land (Exhibit 3.1). The whole of the Yashoda Basin is drained by the Yashoda river and its two tributaries, namely, Bhadadi and Punsoda rivers. The fertility of its soils contribute a number of crops and croplanduse in the study region.

OBJECTIVES OF INVESTIGATION

The prime objective of this study is to assess and analyse geographically, describe and interpret the patterns of agricultural landuse in the Yashoda Basin for the year 1985-86 (Wardha district) with a view to investigate certain physical and socio-economic variables involved in landuse patterns and provide meaningful investigation for making suitable, proper and efficient landuse to meet demand of food for the increasing population and to map, describe and interpret the variations in agricultural landuse patterns with ranking of crop, crop-combinations and diversifications.

Putting it more clearly, present investigation has undertaken following objectives for detailed and exhaustive study as :

(1) To asses and uncover the profile of physical and socio-economic background of the region in order to understand the influence of these factors *namely* relief, geology, climate, drainage, soil, vegetation-cover, population, occupational structure, irrigation, marketing and transportation.

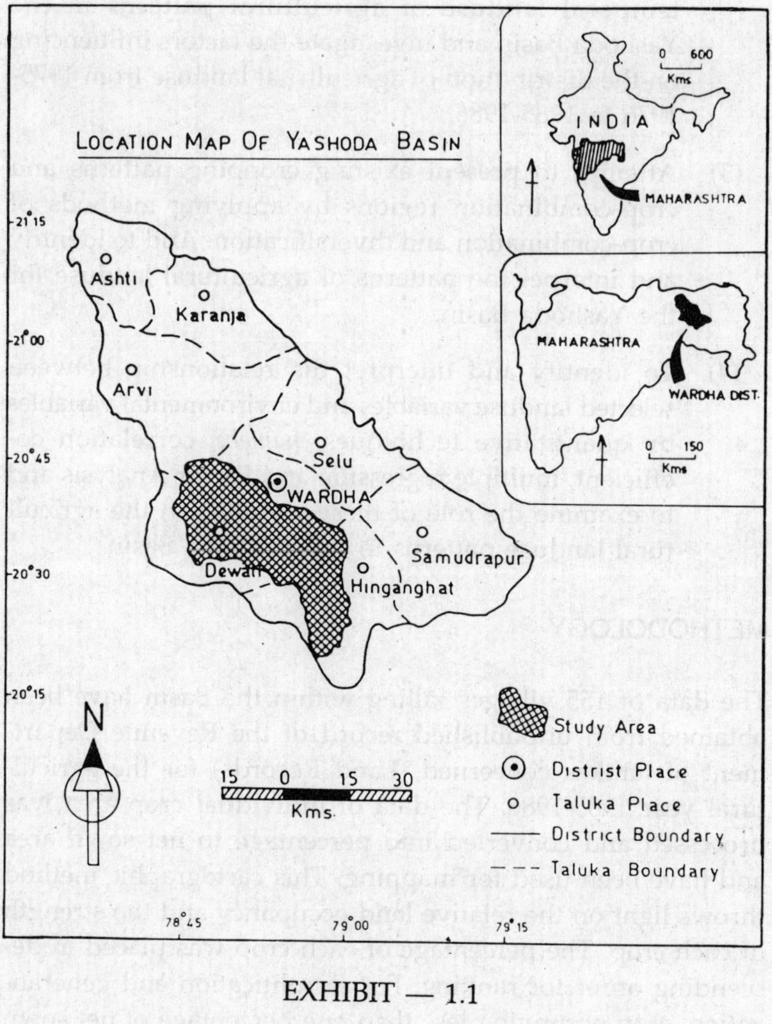
LOCATION MAP OF YASHODA BASIN
INDIA
0 600
Kms
MAHARASHTRA
MAHARASHTRA
WARDHA DIST.
0 150
Kms
21°15
21°00
20°45
20°30
20°15
Ashti
Karanja
Arvi
Selu
WARDHA
Samudrapur
Dewali
Hinganghat
N
15 0 15 30
Kms.
Study Area
Distrist Place
Taluka Place
District Boundary
Taluka Boundary
78°45
79°00
79°15

EXHIBIT — 1.1

(2) Attempt to assess and examine the spatial as well as temporal landuse of agricultural patterns in the Yashoda Basin and investigate the factors influencing on the distribution of agricultrual landuse from 1975-1976 to 1985-1986.

(3) Attempt to present existing cropping patterns and crop-combination regions by applying methods of crop-combination and diversification. And to identify and interpet the patterns of agricultural landuse for the Yashoda Basin.

(4) To identify and interpret the relationship between selected landuse variables and environmental variables by quantitative techniques, *namely*, correlation co-efficient, multiple regression and Factor Analysis and to examine the role of these variables in the agricultural landuse patterns in the Yadhoda Basin.

METHODOLOGY

The data of 155 villages falling within the Basin have been obtained from unpublished record of the Revenue Department of talukas concerned (Land Records) for the agricultural year 1985-1986. The data of individual crop area was processed and converted into percentage to net sown area and have been used for mapping. This cartographic method throws light on the relative land occupancy and the strength of each crop. The percentage of each crop was placed in descending order for ranking. For simplification and generalization, crop occupying less than one percentage of net sown area have been omitted totally in this study. The following statistical procedures have been adopted in the present study :

(1) Doi's method for identifying dominant relative individual crop-combination.

(2) The correlation co-efficient method used for identifying relationship between selected landuse and environmental variables, and,

(3) The statistical measure of Factor Analysis applied to assess the share of each crop involved in the agricultural landuse pattern and identify agricultural regions.

SOURCES OF DATA

The data collected fall into (a) primary and (b) secondary sources supported by interviews of farmers to cross-check their reliability.

The unaccessible and unpublished data have been generated by designing relative questionnaire. This questionnaire covers crops grown in the farm, mode and methods of agriculture practices, type of irrigation facilities, economic conditions of the farmers, loan facilities. The designed questionnaires were circulated among village officials, namely, Sarpanch, Patwari and farmers. The elicited information through questionnaires helped in the interpretation of the contents. Spot inquiries of the farmers during field work served to uncover the realistic picture of total and general agricultural landuse patterns.

The secondary data includes all published materials. It also includes unpublished records preserved in land revenue department. The taluka office was the prime source for area under individual crops for 1985-1986 of the village level. Besides this, published records and abstracts such as, Socio-Economic Review of Wardha District, District Census Handbook, District Statistical Report of Wardha District, Crop and Season Report, Wardha District Gazetteer, Agricultural Bulletin published by Agriculture Department, Maharashtra State, Pune. Periodicals and other documents have been tapped for obtaining relevant data and information bearing on the theme.

The delimination of the Yashoda basin has been attempted by marking out the watershed of river for which following Indian topographical maps (Scale 1:50,000) published by Survey of India are used:

$$55\,\frac{L}{5},\ 55\,\frac{L}{6},\ 55\,\frac{L}{7},\ 55\,\frac{L}{9},\ 55\,\frac{L}{10},\ \text{and}\ 55\,\frac{L}{11}.$$

As the indepth study of all the 155 villages is obviously time-consuming and strenouous work, the choice has fallen on six villages. These villages were selected by systematic purposive sampling method and studied at micro-level. The crop of each field holding has been plotted on the map of sample villages by survey in person. These sample villages may be considered as the representative character, mirroring up the agricultural scenario in the selected year.

The relevant and readily accessible data and information have been collected by preparing questionnaires. The format of questionnaire has been set out. The questionnaires incorporate all aspects of agricultural practices, general landuse, existing cropping pattern, yield of each crop per hectare, various means and intensity of irrigation, socio-economic conditions of farmers, problems and prospects of agriculture.

DESIGN OF WORK

The present investigation consists of ten chapters. The opening *Chapter 1* deals with the background of proposed study covering its objectives, sources of data, methodology used in processing them, analysis and interpretation problems encountered, limitations restricting the scope of this study. *Chapter 2* atempts to present review of relevant literature on landuse studies in India and abroad. *Chapter 3* dwells on the physical environment of the basin. It includes its relief, climate, soils at some length. *Chapter 4* unfolds the socio-economic profile of the Basin, delineating transportation, irrigation, population, and marketing facilities. In *Chapter 5* the investigator has presented the nature and features of the general landuse pattern of Yashoda Basin covering net sown area, forest, cultivable waste, fallow land and land not available for cultivation. *Chapter 6* constitutes the agricultural landuse pattern of all crops grown in the area undertaken for study. This includes cotton, wheat, hybrid jowar, loal jowar, linseed, sesamum, groundnut, ram, mug, mot, and other crops. *Chapter 7* dwells on mere regionalizations of crops grown in the study regions by crop ranking, crop combination and diversification. *Chapter 8* espouses relationship of selected landuse variables and environmental variables. The significance of these variables has been determined by the use of statistical methods, namely, Correlation-co-efficient, Multiple regression and Factor Analysis. *Chapter 9* of this thesis epitomises the intensive and detailed study of the six sample villages, namely, Kachangaon, Sarur, Nagazari, Kurzadi, Rohankheda and Amboda. In this chapter, the investigator has expounded the spatio-temporal analysis of agricultural landuse as far as possible comprehensively and finally, concluding *Chapter 10* presents the resume based on the preceeding indepth and meaningful exposition, analysis and interpretation of the factual informations and data bearing on the whole theme. Papers published and accepted for publication on the present research topic and general references have been listed at the end of this academic exercise. This investigation on agricultural landuse pattern shows up its four levels of study as under :

(1) *Very Generalized Level* - The percentage of net sown area, forest - cover, cultivable waste, fallow land, and area not available for cultivation.

(2) *Extensive Level* - Agricultural landuse of all crops in the Yashoda Basin has been investigated. The crops in the area under study include Jowar (Sorghum Valgare), Wheat (Triticum Spp.), Cotton (Gossypium Spp.), Gram (Cicer Arietinum Linn.), Sesamum (Sesamum Indicum Linn), Mot (Phascolus aconitifolius Jack), Mug (Phascolus radiatuse), Rice (Oryza Sativa Linn.), Groundnut (Arachis Hupogaea Linn.) and Linseed (Linum Usitatissimum Linn.).

(3) *Quantative Level* - This study includes the correlation analysis of selected landuse variables and environmental variables and an attempt is made to regionalize by applying Factor Analysis technique, and,

(4) *Intensive Level* - The micro-study of six smaple villages mentioned earlier are made by employing both temporal and spatial analysis of agricultural landuse pattern for 1988-1989.

PROBLEMS AND LIMITATIONS

The present study converges upon available data for crops at village level in order to depict clear picture of the existing agricultural landuse pattern in the Yashoda Basin. Both accessible and available data and information have been used for analytical, incisive and meaningful analysis. But frequent changes of talukas' administrative boundaries, lack of updated records of crops hectarage at village level have lead to unforeseen problems for the investigation while engaged in field work and data collection from Government offices and while analysing data unanticipated problems alluded like transfers of village patwari that handicapped in getting accurate information and data of crops in concerned village. In many cases, farmers felt apprehensive and reluctant in furnishing correct and reliable data which do not match the published data.

The updated data and information in fulfilling objectives was not readily available hence investigator has confined present study on agricultural landuse pattern for single year and more stress has been given on spatial analysis. There is hardly any access to the data on livestocks and land holdings at village level. Hence these aspects have not been fully elaborated in present study. These constraints and limitations of non-availability of accurate and reliable data spells out handicapped realistic landuse analysis.

2

Review of Literature

INTRODUCTION

The preceeding chapter provides the introductory framework for the present investigation to assess and evaluate the agricultural landuse patterns in the Yashoda Basin for 1985-86 in Wardha District. The study of landuse patterns explicitly relate to the evaluation of land and its characterisics. The main objective of such evaluation is to uncover past and present agricultural landuse patterns. This can serve to reveal improved and suitable landuse planning for the region. In this direction, efforts have been made in India and abroad by many government and non-government agencies where many scholars and planners devoted to expound the land and its characteristics by carrying out pilot landuse surveys. These land use surveys have great significance for further suitable landuse studies. With taking into account the importance of this aspect, investigator, in the present *chapter*, attempted to present brief review of literature related to landuse studies.

CONCEPTUAL FRAMEWORK OF LANDUSE

The concept of regional survey of landuse and its mapping was firstly evolved by Patrick Geddes which had supported by providing concrete framework in World Landuse Survey by Volkenberg (1949). But practical work on landuse study was carried out in 1930 by Late L.D. Stamp in Britain. This is perhaps, the first landuse survey which was, later on, adopted by many Geographers, Economists and planners for landuse studies. Among them, notable work is done by Baker (1937), Agricultural Department in United States (1935), Buck (1929), Shafi (1960), Mohmmad Ali (1967), Ganguli (1964), Karimi (1950) and Lahiri (1950).

WORLD LANDUSE CLASSIFICATION

The landuse studies in India and abroad have been carried out by many scholars in order to understand the land and its characteristics for suitable and proper planning of land in future for increasing high yield per hectare and total production. From this point of view many countries have carried out landuse survey which can be studied as :

(1) World Landuse Studies

(2) Landuse Studies in Britain, America, China and Poland

(3) Landuse Studies in other countries, *namely*, Japan, Cyprus, Bangladesh and Italy

(4) Landuse Studies in India, and,

(5) Landuse Studies in Maharashtra.

World Landuse Studies

The concept and plan of World Landuse Survey was brought forward by S. Van Volkenberg, Clark Unviersity, Massachussetts. He proposed the World Landuse Commission in order to map the World Landuse at the International Geographical Congress at Lisbon in April 1949. The UNESCO supported this idea and appointed a commission in December, 1949 at Clark University to draw up a scheme on the objectives and procedures of the World Landuse Survey. It was proposed that landuse publication may be undertaken on a uniform scale of 1:1 million and each map sheet be acompanied by an explanation text.

In 1953, under the chairmanship of Van Volkenberg at the International Geographical Congress proposed to carry out landuse survey in many parts of the world as possible. It was further decided to divide the world into two regions and Van Volkenberg was Incharge of Americas and L.D. Stamp was placed as Incharge of countries of Eastern Hemisphere.

The first objective of the intented survey was to record the existing landuse in all parts of the World on uniform system of classification and notation with such amplification as may be necessary locally and secondly, the publication of the results including maps and memories. The publication of the detailed maps would only be published by the countries concerned.

The propsed land classification by World Landuse Survey Commission is cited below :

1. Horticulture : It includes the intensive cultivation of vegetables and small fruits
2. Settlement and associated non-agricultural land
3. Trees and perennial crops
4. Crop-land
   (*i*) Continual and rotation cropping
   (*ii*) Land rotation.
5. Improved permanent pasture
6. Unimproved grazing land
7. Woodland, and
8. Swamps and marshes.

On behalf of the commission, worldwide contacts were established on the topic of landuse survey covering Canada,

U.S.A., Japan, India, Pakistan, Bangladesh, Iraq, Burma, Malaysia, Australia, New Zeland, Sudan, Uganda, Kenya, South Africa, Ghana, Nigeria and many other European countries.

Landuse Studies in Four Major Countries

Great Britain, U.S.A., China and Poland have carried out Landuse Surveys which have an immense importance and serve to denote the significant role of landuse studies. The objectives and methods followed were quite different from one another as they have been carried out under varying conditions prevailing in the areas of study, although broadly, all try in their own ways to record the use of land on maps. It would be worthwhile to present a brief outline of these surveys as elaborated below.

Landuse Studies in Britain

The most ambitious and successful landuse survey was carried out by L.D. Stamp in Britain (1930). This survey was executed from 1930 to 1939. The necessary information was collected by organization. The landuse survey scheme was chalked out which, by and large, was applicable to the whole country. It was conducted with the ostensible purpose of making an inventory of what use was then made of every acre of land of Great Britain.

This pioneering work to a large extent was crowned with success with voluntary exercise from the students of schools and colleges of Great Britain. Within less than ten years the endeavour provided a detailed cartographic record of the use to which every acre of land in England, Scotland and Wales has been put.

The bulk of county was surveyed and landuse recorded in the years 1931, 1932 and 1933. The remaining work was finished before the outbreak of Second World War in 1939. The technique of the survey was to make a systematic record of landuse by placing the appropriate letters in its relevant place on the large maps, were, later on, reduced to a scale of 1:63360 and suitable colour was employed to depict different landuse categories. The scale of four miles to one inch (1:253440) or a quarter inch was to the mile was selected for publication of map on account of hard conditions imposed by war.

Landuse Studies in United States of America

The American Land Utilization Survey aimed at improved planning of land for agricultural and other purposes which were generally sponsored. The beginning of land utilization survey in U.S.A. can be traced back to O.E. Baker's published article entitled "Land Utilization in the United States: Geogrphical Aspect of the Problems" (Baker, 1929). He depicted the trends of land utilization and emphasized the need of land classification and surveys.

The concept of landuse planning of country was fostered by United States, Department of Agriculture (Arthur Hillman, 1957). The concept of American Planning can be summarised in a single sentence "What is proposed landuse planning is as broad as country like itself and orgnization should, therefore, presumbly begin with most general functional organization, that can be discovered in an area" (Arther Hillman, 1957).

Three major surveys were conducted in U.S.A. Firstly, a survey was conducted by Tennessee Valley Authority (1936), preceded by Michigan Land Economic Survey (1919) and second, the Rural Land Classification programme of Puerto Rico (1952). Michigan Land Economic Survey, first of these government sponsored project was initiated in 1919 by the Michigan Academy of Science and it started functioning in 1922 with a view to evaluate relative merits of the Northern Michigan in respect of agricultural, recreational or other uses. This survey collected all natural atrtributes of Michigan Land and especially information related to soil, slope, vegetation and similar natural features. The project had been launched, when the wasteland problem in Northern Michigan State became so acute that the whole state began to feel the burden of supporting primary needs of the society (Barnes, 1929). This called for adequate information to base a state sponsored programme to determine the land of Northern Michigan State, best suitable for agriculture, recreation or other uses. In this project, an attempt was made to bring together all agencies capable of compiling a comprehensive inventory.

The second major American Survey, namely, Tennessee Valley Authority was a lineal descendant of the Michigan Project. The technique of "Unit area" was adopted to classify the land of Tennesse Valley by G.D. Hudson. This technique involves "long fraction" and "short fraction" that refer to the method of landuse and its general characteristics, while the denominator indicates physical characteristics of the area. For example, a long fraction with the notation 56322/12113 denotes an area of subsistence farming with major emphasis on grain crops. The cultivated fields of the Tennessee are of the relatively small size, there is limited cultivable land and the farm has moderate level with adequate drainage and slight soil erosion. It has barren outcrop of rocks, soils are deep and modertely fertile. The long fraction serves to depict on the map of a good deal of information on a small areal extent. It followed by short faction which summarises the condition the former reveals. The numerator of this short fraction connotes solely agricultural land use.

Lastly, the Department of Geography, North-Western University, Illinois (U.S.A.) conducted Rural Land Utilization programme of Puerto Rico and collected relative information and mapped it on the scale of 1:10000. This survey was actually started in the year 1949 and completed in the year 1951. The whole island of Puerto Rico was covered by eighteen mapping units. These mapping units varied in areal extents of minimum 323.75 sq kms to 748.5 sq kms. The field work in each mapping unit was done by a field team of three men. Aerial photographs were used to supplement the depicted data in each parcel of land.

The techniques and methods employed in these three American Surveys were suitable for conditions prevailing in the period. The introduction of new technique in the rapidly changing circumstances have become outmoded. The latest surveys employing new techniques have been carried out at both Government and private agencies.

Recent Landuse Investigations in United States

In U.S.A. several agencies were involved in conducting landuse surveys. The investigation about two decades ago was made by Natural and Regional Research Programme. This landuse survey encompassed :

1. Inventories of Landuse;
2. Analysis of development potentials;
3. Examination of compilation and complementary used of land; and,
4. Income and value relationship of different landuses.

The above national landuse inventory had been designed especially to supply data for landuse classifications and water requirements, for enhancing land productions, conservations and management. The relative participation had been active in several national surveys including the following cells, sections of agricultural committee :

1. Land and Water Resources : A Policy Guide, 1962.
2. Basic statistics of soil and water conservation needs 1962.
3. Agricultural land resources, capabilities, uses and conservation needs, 1962.
4. Land and water potential and future requirement for water, 1960, prepared to the Senate Select on National Water Resources.

The organization on National Landuse Inventory conducted landuse studies since 1950, some of these have been completed in eighteen states, namely, Arizona, Delaware, Florida, Iowa, Michigan, Nebraska, Utah, North Carolina, Ohio, Pennsylvania, Virginia and New England State. This study includes use of selected sample counties, sample districts and regional surveys. Moreover, various studies have been made on the basis of random and systematic sampling methods, the use of aerial photo for these studies and aerial photo analysis have provided data with the minimum requirement of trained people and duration of time.

Landuse Studies in China

The landuse study in China was carried out by John Lossing Buck (1929). This has been a unique and fundamental contribution in this field of study. The purpose of Buck's survey was threefold. First, to train learners in the method of research in landutilization, second, to make available knowledge of China's agricultural patterns and features and lastly, to make available the landuse to the people. Who are interested in China's welfare and, lastly, certain elementary information about the state of landuse, food and population of the country.

The sampling technique was employed in survey and village was taken as a representative unit of the region. The survey has proved to be of great value for the collection of data relating to population, food, standard of living and manufacturing. This data collection in 18 locations which were part of 154 hsiens lying in 22 provinces and a bird's eye view of landuse of China has been registered.

Landuse Studies in Poland

In 1953, Professor K. Dziewonsk and J. Kostrowicki have conducted landuse survey in Poland on the guideline of landuse survey done in Britain by L.D. Stamp. The work of land utilization survey was further taken up in 1956 by Agricultural Department directed by J. Kostrowicki. In the years 1956-1959 under the auspices of Institute of Geography of Polish Academy of Science the survey was conducted in 15 counties and plotted on maps sale of 1:100,000 and later on published in the country. This survey was comparatively realistic as it covers arable land, forest types and trees. In this survey, the obvious stress is on the system and direction of utilization of arable land.

Landuse Studies in Five Countries

Japan, Cyprus, Nepal, Bangladesh and Italy have also conducted landuse surveys in their countries. Japan carried out more enthusistically landuse survey of the country in order to prepare uses of land. It is not surprising that landuse survey has received highest priority (Stamp, 1964).

Land Utilization Survey was made by P.P. Karan and prepared maps using aerial photographs. Other countries have also been engaged in conducting landuse studies. Cyprus and Bangladesh are major countries which have carried out landuse surveys. Land utilization maps of Cyprus have been prepared under the supervision of R.R. Rowson and K.R. Selay in the Department of Geography, London School of Economics. The maps of Italy are later on published by Italian National Research Society.

Landuse Studies in India

Geographers have studied land utilization problems in India. Prof. M. Shafi is one of the leading geographer who has made pioneer work adopting sampling method. Dr. L.D. Stamp was in India to attend Science Congress Association at Calcutta (1938). He put impetus on mapping of land utilization and suggested to fallow the model of land utilization survey of England for India. Dr. S.P. Chatterjee (1940) pointed out the necessity of understanding landuse survey in India in the session of Indian Science Congress held at Madras (1940).

The first credit, in India for making an intensive landuse study goes to Prof. M. Shafi. He carried out sample study of land utilization of twelve villages in Uttar Pradesh and relative findings have appeared in "The Geographer" (1951). Prof. M. Shafi presented a most valuable paper to the 20th International Geographical Congress at London (1966) with suggestions on the technique suitable for conducting landuse survey in India. After a brief review of various techniques adopted in different countries recommended that any technique adopted for landuse survey in India should aim at recording existing use of land in the first instance, followed by mapping of land capability or land potentiality at the next stage.

In the years 1965 to 1966 Prof. M. Shafi contributed many valuable papers, namely : (*1*) pattern of croplanduse in the Ganga Yamuna Doab (2) the measurement of food production efficiency in India (*3*) the techniqque of rural landuse planning in India. He published his research findings in the well-known paper entitled "The problems of wasteland in India" in which he uncovers different types of wasteland and suggested methods of reclamations. In Science Today" M. Shafi (1969) concludes that carrying capacity of the Indian land is considered. It can feed five times the present Indian population if it is coaxed well.

Several Universities in India like Aligarh University, Banaras Hindu University, Agra University, Sagar University, have carried out similar programmes either as a part of project work as a doctoral work on landuse studies in their respective universities.

The landuse study programme has been extensively carried out in Aligarh University, Aligarh (Uttar Pradesh). This landuse study was related to the landutilization, nutrition and deficiency as doctoral work under supervision of Prof. M. Shafi since 1962. The following are representative output of research efforts :

1. Geographic basis of Agricultural Landuse in Bundelkhand (1962).
2. Land utilization in the Upper Yamuna Doab (1963).

3. Land utilization in the Central Ganga Yamuna Doab (1964).
4. Land utilization in the District of Pilibhit, Bareilly and Rampur (1965).
5. The Land utilization in the Central Ganga-Ghagara Doab, comprising the District of Faizabad and Sultanpur (1965).
6. The land utilization in the Central Gujarat Plain (1967).
7. Geographic Basis of Agricultural Landuse in the Trans-Rapti plain (1970).
8. The Land utilization in Shahajahanpur and Badaun Districts of Uttar Pradesh (1970).
9. Agricultural landuse in Agra and Mathura Districts (1965).
10. Agricultural landuse and land deficiency in Kumaun region (1965).
11. The land utilization in the Districts of Kheri, Sitapur and Barbanki (1975).
12. Agricultural landuse in the Kashmir Valley (1959).
13. Land utilization in the Ghaggar Plain (1969).

The Department of Geography, Banaras Hindu University, Banaras has studied landuse pattern as a part of ongoing research work. The major important work may be cited below :

1. The land utilization in Chakia Tahsil of Banaras District, Uttar Pradesh (1963).
2. Use and misuse of land in the Kaval towns of U.P. (1966).
3. Landuse in the Lower Middle Gomati Valley (1965).
4. Recent changes in landuse in the Tarai Region in Uttar Pradesh (1960).

5. Problems and patterns of landuse in the Ganga-Ghagra Doab East. (1965).
6. Land utilization in the Neighbourhood of Mirzapur, Uttar Pradesh (1962).

The University of Agra has carried out following work as follows :

1. Land utilization survey in Tarai and Bhabar region of Nainital District, (1967).
2. Land utilization in Kentons Doab of Uttar Pradesh (1966).
3. Land utilization in North-Eastern Uttar Pradesh (1960).
4. Land utilization in Dong Valley of Nepal (1965).
5. Land utilization in Atrauli Tahsil (1966).
6. Land utilization in Sadabad Tahsil (1967).
7. Land utilization and population of Gorakhpur Division (1955).
8. Land utilization in Jaunsar Bawar (1964).
9. Land utilisation in Kanpur District (1962).

The work on land utilization has been carried out in Sagar University, (Madhya Pradesh), Patna University, Delhi University, Calcutta University and Panjab University. The important work of these universities are cited as below :

1. Patterns of land utilization in the Moradabad District of Uttar Pradesh (1961).
2. Agriculture in West Bengal : A Geographical Analysis (1961).
3. Some aspects of the Agricultural Geography of Panjab Plain (1966).
4. Land utilization surveys in the District of Gonda (1965).

5. Agricultural Geography of the Upper Narmada basin comprising the districts of Jabalpur and Mandla (Madhya Pradesh) 1961.
6. Agricultural Geography of Bundelkhand (1967).
7. The study of landuse in a Canal Irrigated area of Patna District (1961).

A number of geographers conducted concrete case studies on land utilization as project works, dissertations work or research work. Some notable valuable contributions made by U. Sen and Guha (1960), R. Lahiri (1950), S.B. Goswami (1960), A. Sharan (1962), S.M. Karimi (1950), M. Ameen (1956), R. Ellefsen (1960), B.N. Ganguli (1956), B.K. Roy (1961), M.S. Honro (1953), C.B. Joshi (1961), Learmonth (1962), B. Jha (1965), K.Z. Amani (1964 and 1968), B.N. Sinha (1968), Joshbir Singh (1971), L. Prasad (1978), V.L.S. Prakash Rao (1956 and 1962), M. Shafi (1956 and 1966), M.F. Siddiqui (1967 and 1968), A. Ahmad (1959, 1967 and 1968), M. Raza (1968), S.N. Mukerji (1968), S.N. Mishra (1964), N.P. Ayyar and S. Srivastava (1968). M.F. Siddiqui (1967), and A. Ahmad (1967), B.G. Tamaskar, Micro studies of the Sagar Damodhar Plateau (1966).

The paper entitled "Techniques of Rural Landuse Planning with reference to India" by M. Shafi (1966) embodies his opinion that the landuse survey like India can be conducted on the basis of smapling as it is a stupendous, laborious and time consuming task to procure data for all the villages in the country. Moreover, the marked similarity in the landuse pattern of extensive regions comprising of many villages. A sample village would be a practical proposition.

Dr. P. Dayal (1957) studied the Agricultural Geography of Bihar. The detail study of villages in India has been pointed out by S.P. Chatterjee (1953, 1956, and 1965). He has suggested rapid surveys to complete whole India within a reasonable time. His work has been documented under the caption of "Land Utilization in the District 24 parganas in

West Bengal (1945). He discussed the influence of soils and climatic environment on land-utilisation, the pressure of population on land and the category of land utilization. In 1953, a detailed land utilization survey was conducted by S.P. Chatterjee to cover the Howrah District and 1200 landuse maps at the scale of 1: 3960 covering 813 villages. Under the guidance of P. Dayal and A. Sharan, E. Ahmad (1959) conducted surveys of three consecutive Community Development Blocks, *namely*, Bihar Sharif, Noorsarai, and Rahui in Nalanda District. In this survey, landuse has been related to physical elements. According to him, the slope of the land should be considered in preparing the development scheme for Indian villages.

Dr. K.Z. Amani (1968) presented two studies to find out the changes that have taken place during the period of forty years (1926-27 to 1966-67) in agricultural landuse and crop production in Harnarayanpur and Golghari villages in Aligarh District (1968). He indicated the reflection of a long process of interaction between physical and socio-economic factors. According to him, the interaction of existing landuse should be based on a thorough inquiry into various variables that have exercised their impact for a long time.

In 1967, M.S. Siddiqui, and A. Ahmad have studied the croplanduse in the Luni Basin in which they identified the crop combinations and the scheme of regional classification emerged out in this study that can well serve as the basic framework of reference in the future micro level studies in agricultural geography of the area.

The study of land utilization of five villages of Ballia District have been made by B.K. Roy (1961). He has studied the physical conditions, occupational structure and changes in the general landuse pattern by comparing the "Khasra Records of 1882-1883 and 1955-1956 and has revealed the acute pressure of population on land and suggested the facility of irrigation to improve agriculture of the region. Another important work carried out by D.S. Chauhan (1966) on "The studies in Land Utilization of agriculture". He discussed the theme covering scope, concept, principles and pattern of land utilization with relative emphasis on landuse policy to promote national plan of economic growth.

R. Lahiri (1950) has analysed agricultural problem in four selected villages near Jasidi Deoghar and has evaluated the fast deterioration of agriculture. M. Guha (1960) made detailed landuse study of typical villages in the Bankura District. S.B. Goswami (1961) selected villages in the Ranchi Districts. Misiri (1956), has used it as a model for the study of general pattern of agricultural landuse on the central Ranchi plateau. M. Ameen (1956) and E. Ellefseen studied the landuse of Delhi State. A.N. Raina (1971) made an attempt to present basic factors affecting types of landuse in the floating garden blocks of the Kashmir Valley.

The landuse study made by B. Jha (1965) on land capability classification of the Kosi region. Jasbir Singh (1971) dealt with the optimum carrying capacity of the land in the Panjab. B.N. Ganguli (1955) has attempted sample survey in a village named Burki. A.K. Sen (1960) carried out the study of landuse survey of the central Mayurakashi Basin of the Birbhum District. A. Anantapadmanabhan (1956) made an attempt to show the relationship between landuse and rural population density in seven representative talukas of Tamilnadu.

Mishra, S. (1964) in his study of landuse in Khadar and raviness of the lower Middle Gomati Valley has attempted landuse planning for better adjustment for optimum utilization and conservation of National Resources. Das, K. (1970) has concentrated his attempt on landuse of Koshi basin in North Bihar, while Roy, B. (1976) selected the districts of Bhojpur and Rohtas for such a study.

Yadav, J.P.S. (1965) has studied the broad regional variations in agricultural landuse in Rajasthan. L.S. Bhatt (1972) has suggested the concept of planning for village, block and district levels. Indupriya Prasad and Singh, S. (1970) have studied landuse problems of many community development blocks of Bihar. Ataullah (1965) in his study of urban landuse in major towns along the Ganga in Bihar has differentiated urban and rural landuse.

R.B. Mandal (1982) has completed a commendable work on land utilization in which he has discussed natural principles, techniques, modes and role of remote sensing and aerial photography in landuse analysis. Several agencies in India have also contributed to the knowledge of landuse studies. The Agricultural Atlas of India, National Atlas and Census Atlas of various states incorporate maps relating landuse. National Atlas and Thementic Map Organisation, Calcutta has prepared landuse maps at different scale.

Landuse Studies in Maharashtra

In Maharashtra, the works on landuse studies have been carried out by three universities namely, Poona University, Shivaji University and Bombay University, either as part of project works or as doctoral works. This works on landuse studies focussed on the land and its characteristics for western Maharashtra. The notable works are as cited below.

1. Agricultural Landuse in Upper Godavari Basin (1978).
2. Agricultural Landuse of the Western District of upland Maharashtra (1981).
3. Spatial Analysis of Agricultural Landuse in Poona District (1984).
4. Changing Agricultural landuse in Kolhapur District (1980).
5. Irrigation and its impact on Agricultural Landuse in Upper Krishna Basin (1982).
6. Some aspects of Agriculture of Satara District : A Geographical Analysis (1980).
7. Agriculture in the underdeveloped Region : A geographical study (1978).

Besides these, the commandable works on landuse studies have carried out as research work for M.Phil course in Shivaji University and Poona University in Maharashtra. This investigation is perhaps the first attempt to present agricultural landuse study in Vidharbha region by investigator on agricultural landuse in the Yashoda Basin.

3

Physical Setting

INTRODUCTION

Land is the basic resource of human society. Its utilization shows a reciprocal relationship between ecological conditions of a region and man. Visher (1932) has rightly put forward the theory that a specific field of geography concerns itself with the study of the influence of natural environment on the nature and distribution of man's activities. The varied nature namely, soil, water, climate provides different area with variety of possibilities of development (Hettner, 1947). It is therefore necessary to evaluate the agricultural landuse of the study region needs to unfold the nature of ecology of the Yashoda Basin.

This chapter uncovers the profile of physical background of the region, *namely*, relief, climate, geology, drainage, soil and vegetation.

Simple cartographic method was applied to show soil and rainfall divisions. Topographical sheets cited in *Chapter I* were used to depict the contour and drainage system and demarcated the Yashoda Basin. The data related to temperature and rainfall have been obtained from Regional Meteorology Department, Nagpur. The rainfall figure for three villages, *namely*, Waigaon, Dewali and Sirasgaon (Bazar) were collected from Irrigation and Water Resource Department, Wardha. The soils division of the study region are depicted on the basis of the map of Wardha district (soil map of Wardha District).

LOCATION AND SPATIAL RELATION

The Yashoda basin falls in Wardha District. Wardha was a part of Nagpur District and was separated in 1862 with its headquarter at *Kawatha* and was later on shifted to *Palkwadi* (1932). This developed new city is now known as *Wardha*. The Wardha District being part of Central Berar, was ruled by Khiljis, Toghluks, Imadshahis, Nizams and later by Marathas.

The areal extent of the Yashoda Basin is within Wardha District. The longitudnal and latitudual extent of the region covers 20° . 20' to 20° . 45' North and 78° .45' to 79° .00 East (Exhibit 1.1). The Basin comprises 155 villages of Dewali, Arvi, Hinganghat and Wardha talukas of Wardha district. It has a length of about 24.5 kms from north to south and 14.5 kms from west to east and covers an area of 860.88 sq. kms (86088 hectares) compassing 13.64 per cent area of Wardha District and 0.27 per cent area of Maharashtra State and population of about 114306 persons (1981 census) with an average population denisty of 92 persons per sq. kms. Table 3.1A reveals the area, population and villages in the Yashoda Basin.

Table 3.1
Villages in Yashoda Basin

Name of villages	*Code No.*	*Name of villages*	*Code No.*	*Name of Villages*	*Code No.*	*Name of Villages*	*Code No.*
Adegaon	121	Bhawanpur	79	Dagapur	104	Jamnala	69
Afazalpur	44	Bhiyapur	31	Dewangaon	77	Jamtha	92
Agargaon	100	Bhagwa	149	Dewali	103	Jawalgaon	57
Ajanavati	82	Bhiwapur	70	Dhamangaon	11	Kanchanpur	6
Ajanavati	114	Bhugaon	42	Dhotra	25	Kachangaon	148
Ajagaon	61	Bhojankheda	129	Dhonapur	109	Kanholi	154
Akoli	108	Bidnagazari	2	Dighi	117	Kawatha	85
Aloda	128	Bodad	21	Digdoh	102	Kelapur	28
Alamdoh	142	Bori	7	Dorali	10	Kesalapur	49

Allipur	139	Bopapur	118	Ekpal	115	Khadaki	94
Amala	14	Borgaon	132	Ekurli	83	Krishnapur	123
Amboda	24	Chanki	146	Fattepur	52	Kurzadi	53
Ambapur	78	Chikhali	119	Fattepur	110	Kurzadi	17
Aminpur	55	Chikani	34	Gadegaon	143	Kurla	89
Ashta	68	Chinchala	143	Gadegaon	145	Laxmi-narayanpur	84
Asala	62	Chitki	86	Giroli	135	Loni	99
Ashrafpur	48	Chunhala	54	Hirapur	37	Lonsawali	9
Babapur	134	Dahegaon	19	Hirapur	106	Malegaon	38
Babapur(rt)	46	Dapuri	71	Isapur	106	Malkapur	88
Banpur	47	Dattapur	73	Itala	113	Mankapur	3
Belgaon	98	Deogaon(Ms)	22	Jaitapur	76		
Barasonegaon	8						
Mamdapur	131	Nimsada	137	Selukate	41	Takli	116
Mansawali	151	Nimgaon	29	Sekapur	87	Takli	125
Meghapur	45	Padegaon	35	Sekapur	127	Takli	138
Mendukdoh	140	Padegaon	30	Shampur	32	Thangadeo	58
Mirapur	72	Palasgaon	101	Shindi(kh)	13	Timapur	105
Mund	133	Paloti	39	Shindi(Mgh)	27	Tigaon	15
Muradgaon	95	Pathari	136	Sirsagaon	150	Umari	16
Muradgaon	111	Pawani	144	Sirasgaon	60	Waddha	51
Nagazari	97	Pimpri	93	Siwapur	67	Wallapur	56
Nagazari	1	Raipur	122	Sonegaon	75	Wai	112
Nagthana	26	Rampur	64	Sonegaon	120	Waifad	18
Nandpur	90	Renukapur	5	Sonegaon	96	Wathada	12
Nandora	107	Rohankheda	141	Sonegaon	147	Yarangaon	153
Narayanpur	59	Rotha	20	Sonpeth	81	Yesgaon	91
Narayanpur	43	Salod	36	Sultanpur	74	Waigaon	63
Neri	66	Sarur	136	Taharpur	65	Jamani	92
Neri	23	Sawangi	33	Talegaon	80	Husenpur	4
Nidha	153	Salsura	50	Takli	155		

Note :	Abj	=	Abaji	Knr	=	Khunkar
	Bzr	=	Bazar	Kh	=	Khurd
	Chnj	=	Chanaji	Mgh	=	Meghe
	Dnd	=	Dhanade	Mskn	=	Miskin
	Drn	=	Darane	Rt	=	Rith
	Ft	=	Fort	Mkn	=	Mokashi
	Jjr	=	Janjal	Ks	=	Kose
	Hpr	=	Hirapur	Nm	=	Nipane

Source : District Census Handbook of Wardha (1971 and 1981)

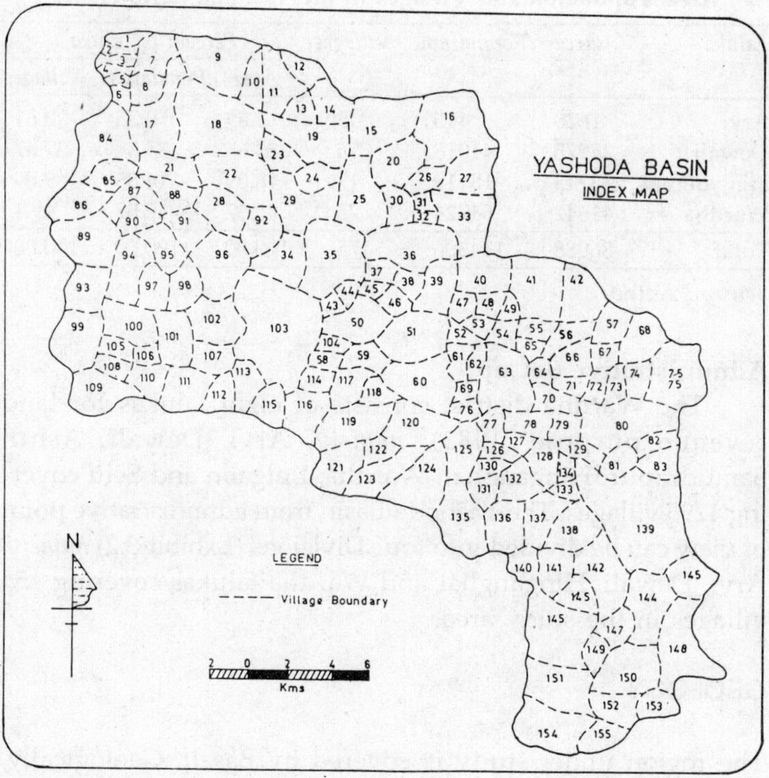

EXHIBIT — 3.1

Table 3.1A
Area, Population and Villages in the Yashoda Basin (1981)

Taluka	*Area*	*Population*	*Villages*	*Percent of taluka*		
				Area	*Population*	*Villages*
Arvi	1628	445	8	1.89	0.38	3.61
Dewali	28975	41018	55	33.65	35.88	37.67
Hinganghat	13743	18119	17	16.09	14.85	9.09
Wardha	41642	54724	75	48.37	47.87	49.1
Total	86088	114306	175	100.00	100.00	100.00

Source : Author

Administrative Set up

The Wardha district consists of eight talukas for land revenue purpose (1983), *namely*, Arvi, Dewali, Ashti, Samudrapur, Hinganghat, Wardha, Pulgaon and Selu covering 1295 villages. The Yashoda Basin from administrative point of view can be divided into four Divisions (Exhibit-3.2) *namely*, Arvi, Dewali, Hinganghat and Wardha talukas covering 155 villages in the study area.

GEOLOGY

The region under study is covered by Basalt. Geologically, the basalt formed by outpouring of enormous lava flow which spread over vast area (Wardha District Gazetteer, 1971). There are two types of basalt, *namely*, vasicular and non-vasicular. The vasicular basalts are soft and suceptible to weather and erosion while non-vasicular basalts are hard, tough, compact and medium to fine grained with conchoidal fracture. Soils in the Basin derived from the rock base duc to constant weathering. The alluvial soil of the region present the transportation of soil particles through course of time. Soil differentiation in the Basin clearly shows the base of two types of basalt and finally reflected on the landuse types in the area under study.

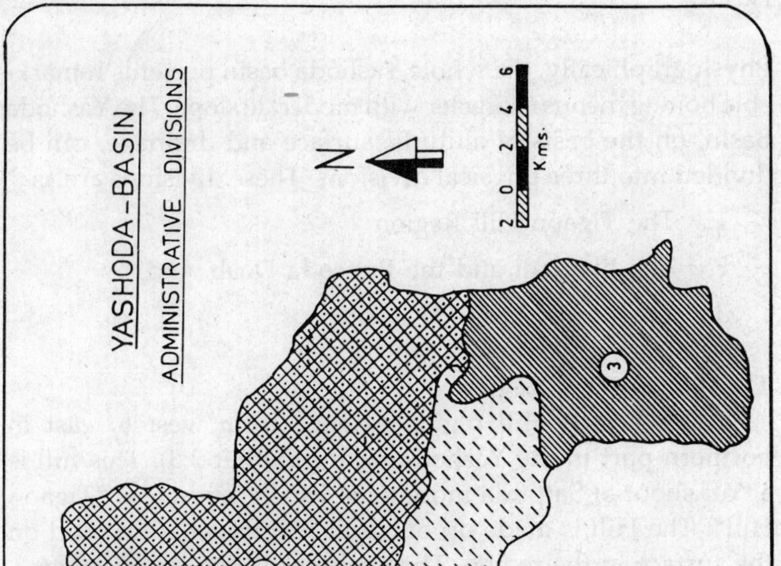
YASHODA-BASIN
ADMINISTRATIVE DIVISIONS
N
2 0 2 4 6
Kms.
1
2
3
4
LEGEND
1 ARVI TALUKA
2 DEWALI TALUKA
3 HINGANGHAT TALUKA
4 WARDHA TALUKA

EXHIBIT — 3.2

RELIEF

Physiographically, the whole Yashoda basin presents remarkable homogeneous character with moderate slope. The Yashoda Basin, on the basis of altitude, surface and drainage, can be divided into three physical divisions. These divisions are as :

1. The Tigaon Hill Region
2. The Bhadadi and the Punsoda Doab, and
3. The Lower Yashoda Basin.

The Tigaon Hill Region

The Tigaon Hill region extends from west to east in northern part in the Yashoda Basin (Exhibit 3.3). This hill is a "off shoot of Satpuda mountain" locally known as "Tigaon Hill". The Hill is made up of basalt and basalt is exposed on the surface in the region. The average altitude is 265 metres. The highest elevation attains 390 metres near Mankapur in the northwest and lowest in the east in study area (200 metres). Rivers rise on this region, *namely,* Yashoda, Bhadadi rivers and numerous small streams. The entire hill region has stony and rugged surface. The unfertile soils of this region can support patchy arable land on slopes giving a small yield of mug, mot, local jawar.

The Bhadadi-Punsoda Doab

The Bhadadi-Punsoda Doab lies in between the Tigaon Hill region and lower Yashoda basin and it is distinguished with identical relief features within sub-regions (Exhibit 3.3). The areal extent of the doab is 47.20 sq. kms. This region is relatively fertile than previous region with greater yield of crops than that of the previous. The gentle slope of the region is also conducive for less run-off. Alluvial soils of this Bhadadi-Pursoda doab are composed of fine grains derived from the eroded surface of Tigaon Hill region. Cotton, wheat and hybrid jawar are successfully grown in the region.

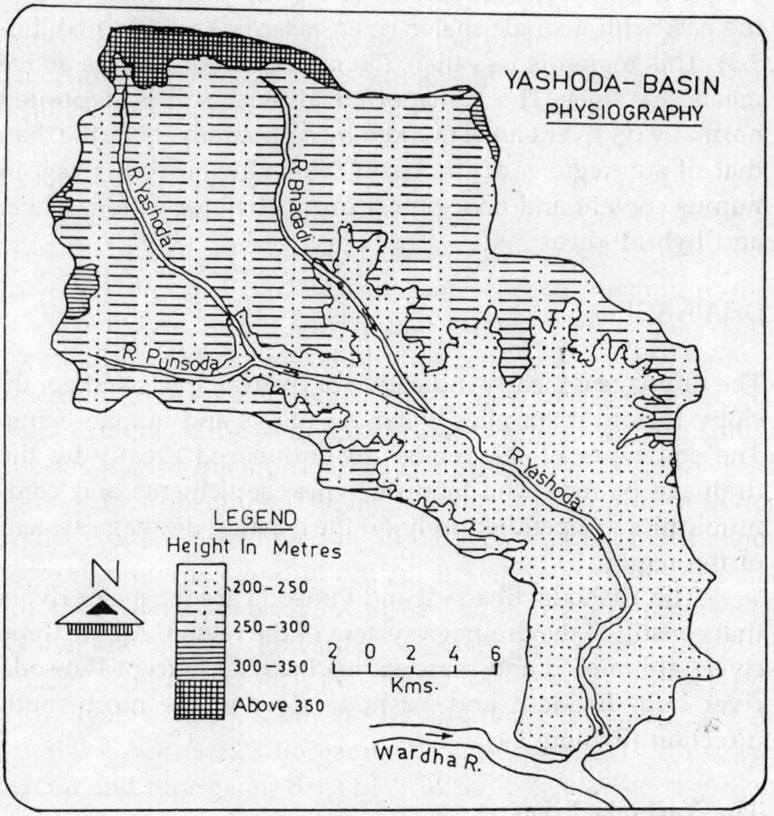
YASHODA- BASIN
PHYSIOGRAPHY
R. Yashoda
R. Bhadadi
R. Punsoda
R. Yashoda
LEGEND
Height In Metres
200-250
250-300
300-350
Above 350
N
2 0 2 4 6
Kms.
Wardha R.

EXHIBIT — 3.3

The Lower Yashoda Basin

The lower Yashoda Basin covers 45.20 sq. kms. area in the east with a single major river, *namely*, Yashoda (Exhibit-3.3). This region is less than 250 metres in its altitude above mean sea level. The transported alluvial soil is deposited normally on river banks marked by comparatively fertile than that of sub-regions of the basin. This alluvial soil is rich in humus content and hence most productive for cotton, wheat and hybrid jowar.

DRAINAGE

The drainage is clearly indicative of riverine tracts and gently slopy region. It supplies water for plants and human being. The soil types of the region are influenced greatly by the drainage pattern. The landuse types, settlements and communications depend entirely on the existing drainage system of the region.

The Yashoda, Bhadadi and Punsoda are the major rivers that constitute the drainage system of the region. All the three rivers are rainfed and seasonal in character. Except Punsoda river both Bhadadi and Yashoda flow in the north-south direction (Exhibit 3.4).

The Yashoda River

Yashoda, the prime river in the study region is rainfed and seasonal in character. Besides use of water for drinking purposes by settlements, riverine water is used for irrigation. The river Yashoda rises in Tigaon Hill at the height of 370 metres above mean sea level and after flowing in the north-south direction about 8 kms and joins the Wardha river near village Kanholi. The course of river becomes wider along the south. Rapids are the common features in the upper course of the river. The Yashoda in rainy season brings fine grains and clays and deposits them in lower Yashoda basin and forms alluvial tract where varied crops are grown.

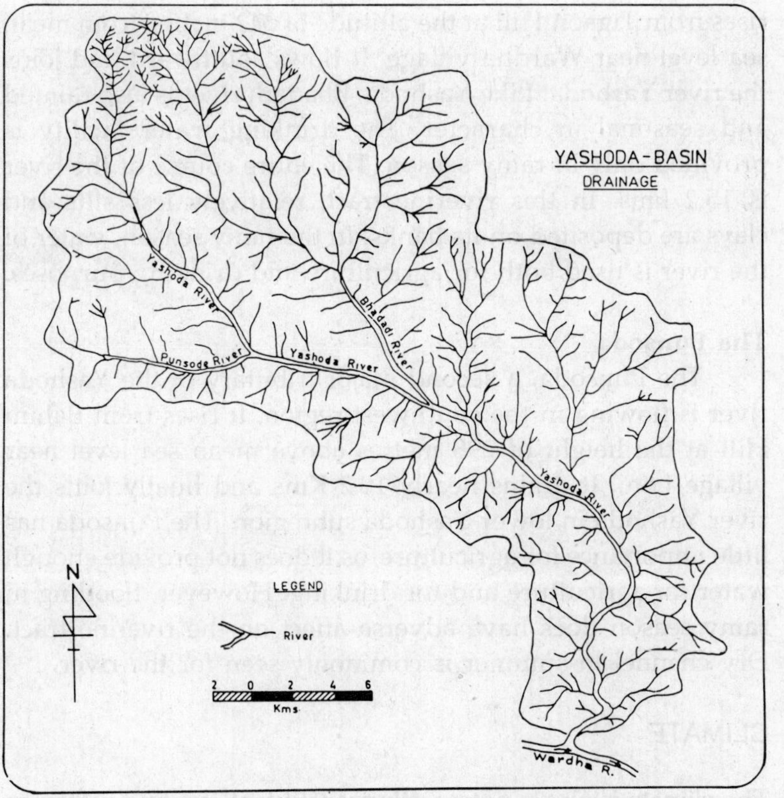
YASHODA-BASIN
DRAINAGE
Yashoda River
Bhadadi River
Punsode River
Yashoda River
Yashoda River
N
LEGEND
River
2 0 2 4 6
Kms
Wardha R.

EXHIBIT — 3.4

The Bhadadi

The Bhadadi, a major tributary of the Yashoda river also rises from Tigaon Hill at the altitude of 342 metres from mean sea level near Wardha village. It flows southward and joins the river Yashoda. Like Yashoda, Bhadadi river is also rainfed and seasonal in character. The drinking water facility is provided only in rainy season. The entire course of the river is 15.2 kms. In this riverine tract, relatively less silts and clays are deposited on its banks. In the rainy season, water of the river is used both for agriculture and drinking purposes.

The Punsoda

The Punsoda, a second major tributary of the Yashoda river is flowing in the southwest region. It rises from Bahim Hill at the height of 298 metres above mean sea level near village Loni. It drains nearly 10.2 Kms and finally joins the river Yashoda in lower Yashoda subregion. The Punsoda has little importance for agriculture, as it does not provide enough water for agriculture and for drinking. However, flooding in rainy season does have adverse affect on the riverine tract. Dry channel in summer is commonly seen for the river.

CLIMATE

The climate is also a noteworthy constituent exerting influence on crops in the Yashoda Basin. The vegetation cover and the prevelant climate are the variables entering into the formation of soils and this soil rises to give for existing agricultural landuse. The proper and suitable climate is essential for better yield, growth, production and reproduction (Chauhan 1971).

The climate in the Yashoda basin is typical monsoonal in character with a small range of rainfall and temperature. The study region experiences intensive heat in summer and intensive cold in winter. The intensive heat in summer is more dry and affects the vegetation in shunting its growth.

The Yashoda Basin climatically can be divided into four

seasons as below :

1. The cold winter season.
2. The hot weather season.
3. The rainy season, and
4. The post-monsoon season.

The Cold Winter Season

The cold winter season is restricted to four months, *namely*, November, December, January and February. It commences in mid November and ends in February. Clear sky, fine weather, low humidity and large durinal variations in temperature are common characteristic features in cold winter in Yashoda basin. The temperature starts to go down from November and it continues further in February. Both day and night temperature decrease fairly and rapidly with the advent of winter. Table 3.2 shows the average monthly rainfall and temperature at Wardha. In cold season rainfall amount is negligible in the study area. Temperature experiences during these months are dropped down considerably.

Table 3.2
Monthly Average Temperature at Wardha

Months	*Temperature (in °C)*	*Rainfall (in mms)*
January	20.90	22.13
February	24.90	16.76
March	22.40	Nil
April	30.50	1.70
May	30.80	20.66
June	30.10	215.20
July	30.00	341.53
August	27.10	274.06
September	26.70	173.13
October	26.80	37.80
November	22.80	2.08
December	21.00	Nil
Annual Average	26.33	92.02

Source : Regional Metereology Department, Nagpur.

January appears to be coldest (Table 3.3 and Exhibit 3.5/A) in the year. Maximum temperature of November and December are 32.2°C and 30.8°C respectively. February experiences a temperature of 32.6°C (Table 3.3).

The winter season with usually dry weather has a beneficial effect on the winter crops, *namely*, wheat. The occasional stormy rain in the season harms the standing crop in the winter. The standing crops are swept down on the soils.

The Hot Weather Season

March, April, May and first half of June constitute hot weather season in the Yashoda basin. This season is very unpleasant and is dominated by westerly dry winds. The nights are uncomfortable and days are marked by intensive heat. The average temperature in basin during summer is 28.4°C. Right from March, the day temperature starts rising and reaches its maximum in May (Table 3/3, and Exhibit 3.5/A). May is the climax of the year with intensive and unbearable heat (41.8°C temperature) whereas. March and April experience 40.2°C and 40.8°C temperature respectively. Excessive heat has desiccating influence on vegetation. The trees are shading their leaves to protect from intensive heat in summer. The relative humidity in summer remains always less than that of other seasons. The *Andhi* or high speed dry winds or *Tufan* are common phenomena which raises a huge cloud of dust in the sky. Such phenomena originates due to abnormality in weather. The weather in March and April is conducive for ripening, threshing and winnowing of the rabi crops.

THE RAINY SEASON

During the rainy season in the Yashoda basin nearly ninety per cent crops are raised. The regime and amount of rainfall serve to shape the landuse pattern. The early or late outbreak of the monsoon hampers the sowing operations. The mansoon, generally, bursts in the middle of June and lasts till September.

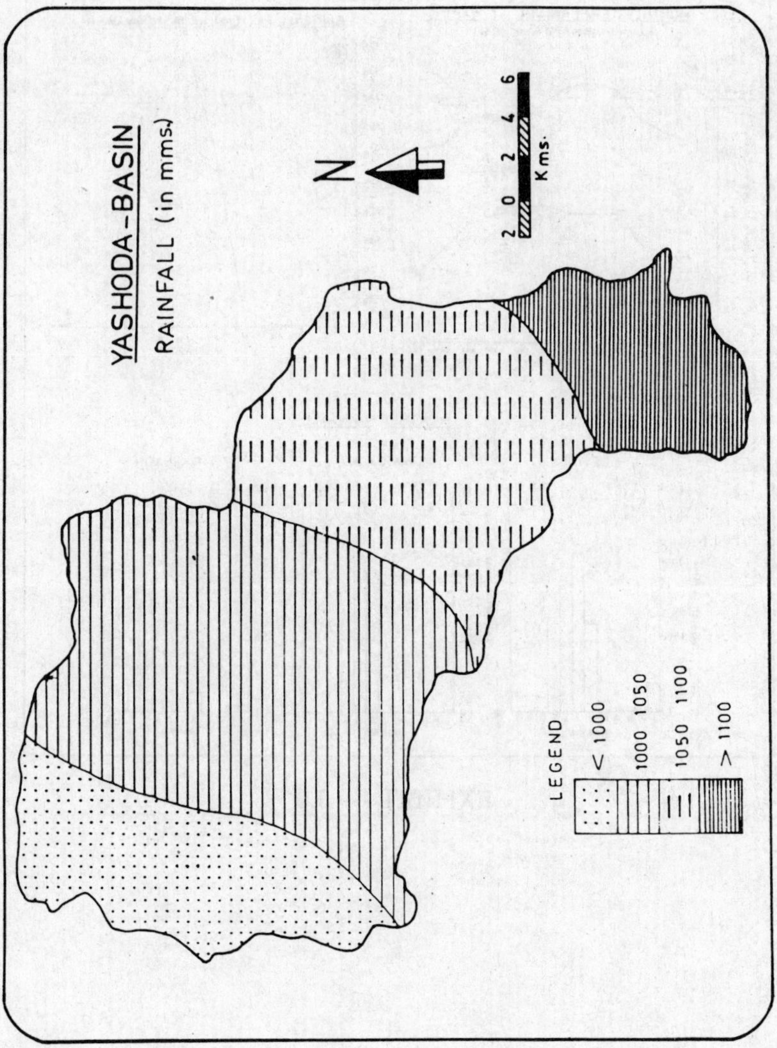
YASHODA-BASIN
RAINFALL (in mms.)
N
2 0 2 4 6
Kms.
LEGEND
< 1000
1000 1050
1050 1100
> 1100

EXHIBIT — 3.5

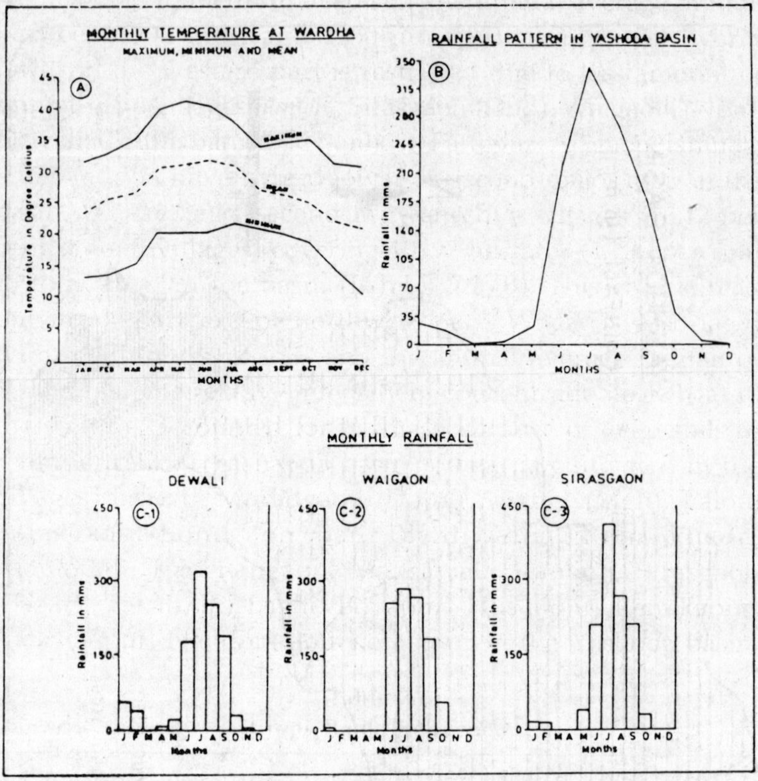
MONTHLY TEMPERATURE AT WARDHA
MAXIMUM, MINIMUM AND MEAN
A
Temperature in Degree Celsius
MONTHS
RAINFALL PATTERN IN YASHODA BASIN
B
Rainfall in mms
MONTHS
MONTHLY RAINFALL
DEWALI
WAIGAON
SIRASGAON
C-1
C-2
C-3
Rainfall in mms
Months

EXHIBIT — 3.6

Table 3.3
Maximum and Minimum Temperature at Wardha

Months	*Maximum Temperature in degree C*	*Minimum Temperature in degree C*
January	32.4	13.4
February	32.6	14.2
March	40.2	14.6
April	40.8	20.2
May	41.8	19.8
June	40.8	20.2
July	38.0	20.0
August	33.0	21.2
September	34.0	19.4
October	35.2	18.4
November	32.2	13.4
December	30.8	11.2

Source : Regional Metereology Department, Nagpur.

Table 3.4
Monthly Average Rainfall at Three Stations in Yashoda Basin

Months	*Rainfall station (Railfall in mms)*		
	Dewali	*Waigaon*	*Sirasgaon (Bazar)*
January	52.00	9.00	14.40
February	37.20	5.00	13.10
March	Nil	Nil	Nil
April	5.00	Nil	Nil
May	24.40	Nil	37.60
June	178.00	258.60	209.60
July	320.00	287.20	417.40
August	234.00	270.20	268.00
September	188.40	190.60	210.40
October	30.00	55.40	28.00
November	2.00	Nil	4.00
December	Nil	Nil	Nil
Total	1071.20	1076.00	1202.50

Source : Water Resource Department, Central Rainguage Water Board Wardha Division, Wardha

The advent of rainy season brings change in weather condition. The restless nights and days of summer are converted into comfortable days. Temperature goes down steadily (Table 3/4 and Exhibit 3.5/A). The distributional pattern of annual rainfall in Yashoda Basin is shown in Exhibit 3.5/B. The Tables 3.5C, 3.5D and 3.5E present the average monthly rainfall in three villages, *namely*, Dewali, Waigaon and Sirasgaon (Bazar). These three stations record rainfall in increasing order with its maximum at Sirasgaon (Bazar) (Exhibit 3.5/C/3) in the Yashoda basin.

This season of the year is characterised by heavy and prolonged rainfall with large scale inflow of maritime air. Uneven distribution of rainfall is common in this period. Cloudless skies, accompanied by long spells of bright sunshine dry out the standing crops. Dry spells of weather intervene in between wet days, which are unpredictably uneven. This phenomenon has its impact on the yield of crops (This has been correlated in detail in *Chapter 8*).

Post-Monsoon Season

This season prevails from October to November due to rapid and largest decreasing amount of rainfall in late September. These two months receive little rainfall. Table 3.3 shows the average rainfall in October (37.8 mms) and November (2.08 mms). Increasing day temperture drops down rapidly in afternoon. The relative humidity is rather reduced than that of rainy season.

HUMIDITY

The air is generally dry over the Basin except in the monsoon season. Before commencing monsoon and therefore, the humidity considerably drops down. The high humidity is observed in the rainy season. It decreases to 30 per cent in December and to less than 20 per cent in May due to extreme dryness. July and August have highest humidity (Wardha District Gazeetter, 1971).

CLOUDS

The Yashoda Basin appears generally to have clear skies throughout the year except during rainy season. During rainy season the clouds overcast the sky. This phenomena of continuous clouds in sky have adverse effect on crops and the cropping patterns. The flowering of tur plants drops down and thus adversely affects the crop.

WINDS

Strong wind like *Andhis* or *Tufan* in summer have adverse effect on standing crops in the basin. Strong winds in rainy season uproot plants of crops. But most period of the year remain unaffected as the harvesting is over earlier.

SOILS

The soils in the Yashoda basin is the key parameter influence on the existing cropping pattern. Soils vary with changing climatic condition in region. Soils and its inherent properties give control on the crops and crop-yielding. Moreover, its mineral composition, texture, thickness and fertility set fundamental limit to the landuse pattern. The fertility of land depends on organic, inorganic material and water available in soil. It is, therefore, necessary to make a detailed study of the soil for proper landuse (The association of crops and soils have been discussed in Chapter 6). Jainedra Kumar (1985) has rightly pointed out the need for careful study of soils in order to make efficient landuse as soil provides basic nutrients to plants far longer than chemical fertilizers. The significance of soils can be assessed by the following facts :

(*1*) It provides physical support to vegetation

(2) It serves as a medium for storing water and bringing water into contact with roots, and,

(3) It supplies a small but essential percentage of material which are converted into food by plants.

The assessment of soil of Yashoda basin has not so far been made earlier. The soil divisions map of Yashoda basin was prepared on the basis of soil map supplied by Agriculture Department, Pune. It is broad soil divisions of the Wardha district.

Table 3.5
Soil Types in the Yashoda Basin

Soil Texture Classes	*Proportion in percent*				*pH*	*Soil Colour*
	Coarse sand	*Fine sand*	*Silt*	*Clay*		
Clay	11.90	22.08	13.75	48.00	8.5	Dark Black soil
Sandy clay	15.85	33.03	06.25	41.75	8.0	Dark Black soil
Clay Loam	17.30	09.70	28.75	38.00	7.8	Medium Black soil
Sandy clay Loam	15.00	30.76	20.00	28.00	7.7	Medium Black soil
Sandy Loam	26.70	29.55	22.50	18.50	8.3	Shallow coarse soil
Silty Loam	13.70	13.23	52.50	18.00	8.1	Shallow soil

Note : The above analysis was carried out in the Soil Testing Laboratory, Department of Agriculture Pune - 411005.

This map does not give realistic picture regarding colour, texture, slope and drainage. Therefore, spot-inquiries and soil samples were the main source of soil interpretation for Yashoda basin. Majority of soil types of the basin are formed *in situ* by weathering process of the prevelant basalt. It is rich in nutrients such as lime, iron and alkalies (Gazetteer, Wardha District, 1971). The information obtained from soil analysis of samples have been incorporated in Table 3.5.

On the basis of chemical constituents, shades of colour and texture, soils fall into three types as follows :

(1) The coarse shallow soil,

(2) The medium black soil, and

(3) The deep black soil.

(1) *The Deep Black Soils*: This soil covers 6.3 per cent of the study area lying in lower Yashoda basin along the river Yashoda (Exhibit 3.7). The black soil in region is called "Kali" or "Morad" or "Regur". It appears as dark brown to grayiesh black in colour. The soil particles are very fine and coherent. It is black in colour due to excessive predominance of humus content over ingradient and has high power of retaining moisture. It becomes pasty in rainy season. This soil contains 0.35 per cent nitrogen 1.28 per cent potash and 14 per cent calcium, whereas pH is 8.2 (Field work and soil analysis).

The collected soil samples during field work have been analysed and results obtained are set fourth in Table 3.6.

Table 3.6
Analysis of Deep Black Soil

Soil types	*Soil*	*Parameters (in %)*			
Coarse sand	17.80	N	0.35	CaCo3	4.40
Fine sand	22.20	K	14.00	Co	41.58
Silt	13.75	P	1.28	Mg	8.17
Clay	43.00	pH	8.2		

Note : The above analysis was carried out in the Soil Testing Laboratory, Department of Agriculture, Pune - 411105.

The deep black soil in the Yashoda basin varies in depth from two feet to ten feet. The ploughing of the soil in summer is not easy as it becames very hard. It develops cracks in summer season and sometimes such cracks in summer season appear more than three feet deep. The deep black soils of this region have good fertility due to its high moisture retentive capacity. This soil appears in the villages, namely, Nidha, Mansawali, Gadegaon, Rohankheda, Sirasgoan

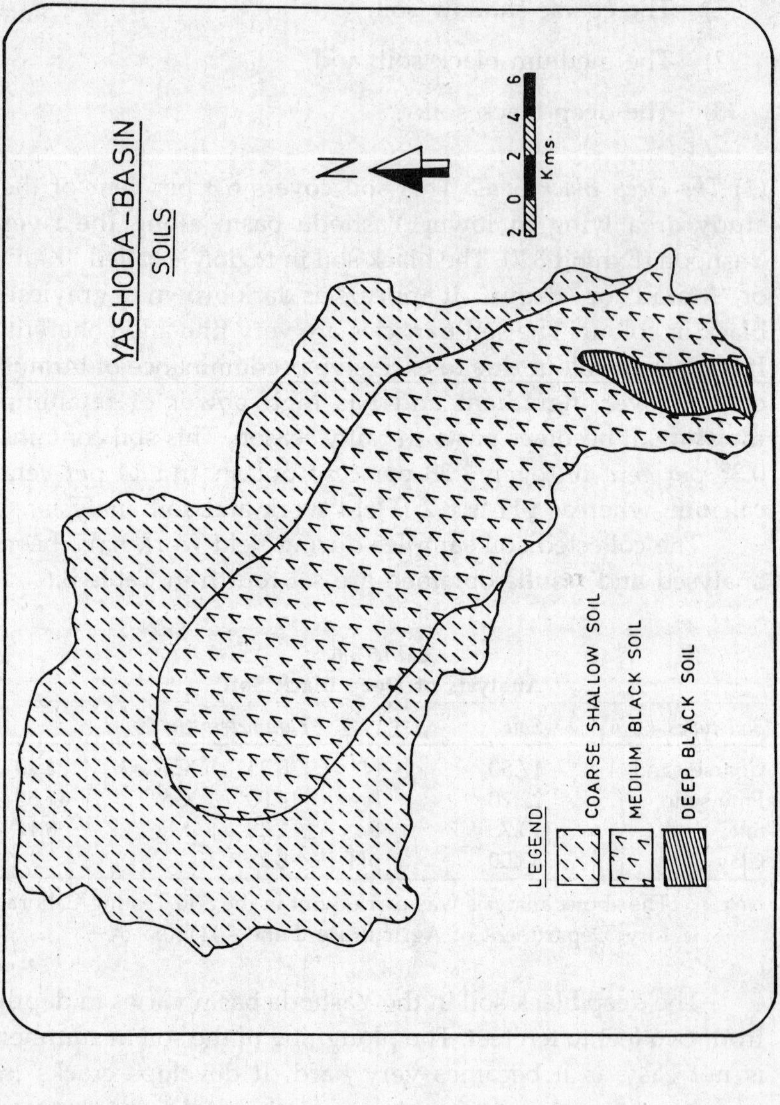
YASHODA-BASIN
SOILS
N
2 0 2 4 6
Kms.
LEGEND
COARSE SHALLOW SOIL
MEDIUM BLACK SOIL
DEEP BLACK SOIL

EXHIBIT — 3.7

(Bazar), Kanholi, Yarangaon and Pawani. Though all the crops can be cultivated on black soil, cotton and hybrid jowar are predominant crops in this soil type as this soil has high yield per hectare.

(2) *Medium Black Soils*: This soil covers 46.20 per cent area in the basin (Exhibit 3.7). It is relatively less fertile than earlier soil type as the proportion of sand and gravel increases in medium soil. This soil can be characterised by less clod and less fine mud is saturated in dry period. Locally, this soil is known s "Baradi Land" as it combines approximately 50 per cent sand and clay. Table 3.7 shows analysis of medium black soil in the Yashoda basin.

Table 3.7
Analysis of Medium Black Soil

Soil types	*Soil*	*Parameters (10%)*			
Coarse sand	15.10	N	0.27	CaCo3	8.7
Fine sand	34.85	P	1.72	Ca	43.20
Silt	11.60	K	20.50	Mg	10.32
Clay	33.00	pH	7.8		

Note : The above analysis was carried out in Soil Testing Laboratory, Department of Agriculture, Pune 411105.

The medium black soil acquires loose particles in dry and muddy in wet and lime content occurs in a moderate quality. Moreover, its soil soaks water easily and has less productivity than earlier. The cotton, hybrid jowar, local jowar, wheat, tur, rice are raised successfully in this soil. This soil in the region appears in central and lower Yashoda basin.

(3) *The Coarse shallow Soils*: This soil type is confined in the north, west and eastern parts of the basin covering 47.50 per cent of the region (Exhibit 3.7). Locally this soil is known as 'Kharadi Jamin'. The analysis of coarse shallow soil sample is given in Table 3.8.

Table 3.8
Analysis of Coarse Shallow Soil

Soil types	*Soil Parameters (in %)*				
Coarse sand	22.70	N	0.26	Caco3	9.40
Fine sand	29.55	P	0.84	Ca	30.24
Silt	22.50	K	4.73	Mg	15.05
Clay	18.50	pH	8.1		

Note : The above analysis was carried out in the Soil Testing Laboratory, Department of Agriculture, Pune 411105.

This soil type has predominance of coarse sand (22.70 per cent) and silt (22.50 per cent). The particles of sand finds closely spaced, rapidly soaked in nature. This soil has low evaporation rate than earlier soil types and is poor in phosphorus and nitrogen (Table 3.8). All crops can grow but with low yield due to low productivity of the soil. Rice and local jowar are raised in this soil type in the Yashoda basin.

VEGETATION

There is little or no identified forest-cover as defined by the Forest Department (Wardha District Gazetteer, 1971). The scanty and sparse vegetation in the Yashoda basin is confined to hilly and riverine tracts. This vegetation is of mixed and scrub type which drop their leaves in summer in order to protect themselvess from unbearable intensive heat. Along the rivers and small streams have sparse tress, *namely*, Mango (Mangi fera Indica), Jambhul (Syzygium Cumini), Bhabhul (Acacia Arabia), Nimb (Melia Azadarachta), Palas (Butea Frondosa), Pimpal (Ficus Religiosa), Bor (Zizuhus Jujuba, Chinch (Tamarindus indica), Moh (Madhuca Latitolia), Moharukh (Ailanthus excelsa) Sitaphal (Anana Sanamosa), and Katbor (Ziziphus Oenoplia), and weeds like harali, kunda, tunoali, gujar and gorvali are found in the basin. The vegetation of the region does have insignificant effect on agricultural landuse pattern as it covers less and dispersed areal extent in the Basin.

RESUME

The present chapter deals with the profile of physical aspects of the Yashoda basin, namely, relief, drainage, soil types, climate and vegetation cover. The study uncovers the remarkable homogeneous character with moderate slope and plain region except the source of the river Yashoda. The climate in the basin has small amount of rainfall variations and high temperature. January is the coldest month with average monthly temperature of 20.9°C and the month of May records the maximum temperature of 41.0°C centrigrate. Evidently, Basin experiences intense heat in the summer and fairly cold weather in the winter. The Yashoda basin receives rainfall from the southwest monsoon. With the onset of the monsoon, cyclonic storms visit the basin. July and August months witness heavy rainfall (341.53 mms and 274.06 mms) respectively in the basin. This fact is supported by the rainfall recorded at Dewali, Waigaon and Sirasgaon (Bazar) (Exhibit 3.5/C-3). Soil types of the Basin are formed *in situ* by weathering process of prevelant basalt. It is rich in nutrients such as lime, iron and alkalies, scanty and sparse trees are common within the Yashoda basin.

4

Socio-economic Setting

INTRODUCTION

After unfolding the background of physical setting of the Yashoda basin in Chapter 3 partaining to climate, soils, and relief structure, it would be relevant to understand the role of socio-economic aspects in shaping agricultural landuse pattern in area under study. Both the physical and socio-economic variables exert their impact on agriculture practice and its production. The traditional system of agriculture and decision making have their bearing on the regional variations in agricultural landuse in the Yashoda basin. This has been elaborated and discussed in this Chapter. Jasbir Singh and S. Dhillon (1987) have rightly stressed the necessity for the evaluation of socio-economic variables in terms of inputs involved in agriculture sector that have been ultimately forming landuse pattern and yield per hectare. The present study, therefore, is significant in this respect.

The objective of this Chapter is to highlight the socio-economic background of the region, namely, population, occupational structure, irrigation, marketing, transportation and their impact on landuse pattern of the area under study.

The data of population in 1971 and 1981 have been collected from Wardha District Census Handbook, Wardha. Area under irrigation has been obtained from Land Records of

Taluka concerned. The elaboration on land holding is entirely based on spot inquiry and sample survey during the field work. The suitable cartographic methods are used to depict the density of population, occupational structure and irrigation of the basin.

POPULATION

The population is an important resource from economic point of view for the regional development of agriculture as it influences the economic activity and determines the level of consumption and agriculture force. It also forms the workforce, *namely*, cultivators, agricultural labourers.

The population in the basin is 114306 persons (1981 census). It accounts for 12.33 per cent of the Wardha District and about 0.18 per cent in the Maharashtra State. The average density of population in the area under study is 132.77 persons per sq. km while agricultural density is 0.65 persons per hectare. 87.57 per cent working population to total has engaged in agricultural activity and rest are involved in household, industry, trade and transportation in the area under study.

The population data partaining to growth are not readily available for Yashoda basin. Therefore, investigator has obtained the population data from Wardha District Census Handbook and Wardha District Gazetteer. The Table 4.1 shows the population since 1901 with its decadal variations and percentage of growth.

It is observed from the above table that the population in the Yashoda basin has been steadily increasing from 1901 to 1981. The population in basin in 1901 was 40119 while it increased and attained 114306 persons in 1981. The total increase in population during the above period is 74187 persons with higher rate (Table 4.1) excepting the decade in 1911 to 1921. During this decade, spreading (1918-1919) of influenza epidemic coupled with crop failure has decreased the growth of population (3.04 per cent). Moreover, in the period between 1931 and 1951 the growth of population did not increase considerably (7.29 per cent) owing to spreading out Maleria and Cholera which took heavy toll of human life. That is why, the basin shows low rate in the above mentioned period. However, from 1951 onwards the population increased with high rate. It has increased by 17.32 per cent growth in 1971 and 17.99 percent in 1981. The ratio of cultivated land and human power in the basin is 0.59.

Table 4.1
Population Growth and Decade Variations

Years	*Population*	*Decade Variations*	*Growth (In per cent)*
1901	40119	—	—
1911	47151	+ 7032	+ 14.91
1921	48630	+ 1479	+ 3.04
1931	58951	+ 10321	+ 17.50
1941	63593	+ 4642	+ 7.29
1951	68143	+ 4550	+ 6.67
1961	77505	+ 9362	+ 12.07
1971	93741	+ 16236	+ 17.32
1981	114306	+ 20565	+ 17.99
From 1901 to 1981		+ 74187	+ 64.90

Source : Author

Density of Population

This aspect has been studied in order to understand the regional variations in the population density and its influence on agricultural landuse. The area under study has 87.57 per cent working population which is directly involved in agriculture (1981). The population density value was here calculated as ratio of total population to the total area from 1901 to 1981 and is shown in Table 4.2.

It is revealed from the above table that density of population in the basin is increasing since 1901 from 46.6 persons per sq. km to 132.77 in 1981. The interesting fact regarding the density of population increased threefold during eighty years. The decade from 1911 to 1921, the density exhibits low rate (56.48 persons per sq. km.), while high rate of density was observed in decade 1971 to 1981 (132.7 persons per sq. km.) in the basin.

Table 4.2
Denisty of Population (Years 1901 to 1981)

Years	*Population*	*Area (Aq. Kms)*	*Density (Persons per Sq. Kms)*
1901	40119	860.88	46.60
1911	47151	860.88	54.77
1921	48630	860.88	56.48
1931	58951	860.88	68.47
1941	63593	860.88	73.86
1951	68143	860.88	79.15
1961	77505	860.88	90.02
1971	93741	860.88	108.88
1981	114306	860.88	132.77

Source: District Census Handbook and District Gazetteer, Wardha.

The Spatial Distribution of Population Density

Exhibit 4.1 reveals the spatial distribution of population density in the Yashoda basin for 1971 and 1981. It is observed from this exhibit that there is considerable spatial variations in the density distribution within the region. The highest density in basin is 391.52 persons per sq. km. at Dewali in the central part and lowest at Mankapur in the north (11.32 persons per sq. km.). The density of population distribution is increasing towards east in the basin (Exhibit 4.1/A and 4.1/B). More than 300 persons per sq. km. density in 1971 is observed in four villages (2.58 per cent to total villages), *namely*, Waigaon (376.34), Salod (426.03), Sindi (354.34) and Kurzadi (304.56). These villages have high density due to market like Wardha market centre where vegetables, in addition to, cotton and wheat are produced to a great extent. Less than 50 persons per sq. km. density is in the northern and western parts in the basin in the following villages, *namely*, Nagazari (11.32), Bid-nagazari (15.2), Husenpur (13.33), Kanchanpur (13.10), Mankapur (11.39), Nandpur (28.61), Khadaki (49.33) and Pimpri (46.59). This area of low density attributed to less fertile soil of the region having low yield per hectare. The population density group between 200 to 250 persons per sq. km. recorded at Selukate (237.39) and Kawatha (247.30). 50 to 100 persons per sq. km. density is concentrated in central and eastern parts in the ara under study (Exhibit 4.1/A).

The population density pattern in the Yashoda basin for 1981 has slightly changed than 1971 (Exhibit 4.1/B). This exhibit shows the density between 50 - 100 persons per sq. km. has been replaced by 100 to 150 persons density group. In eastern part of the basin, the population density occurs from 200 to 250 persons per sq. km. The villages included in this category are, Allipur (230.07) and Talegaon (212.38). The northern and western parts in the basin, villages have low density owing less area extent under crop being less fertility. The villages, *namely*, Nagazari (11.90), Bid-nagazari (15.17), Kanchanpur (14.11) and Mankapur (11.37). The Selukate in east of the basin observes 275.10 persons per sq. km. population density (Exhibit 4.1/B).

OCCUPATIONAL STRUCTURE

The availability of labour resource and its involvement in various activities in agriculture represent the scenario of development of the region. The Yashoda basin has preponderance of agriculture, engaging 87.57 per cent population as working force (1981). The population can generally, be grouped into two groups (*a*) working population and (*b*) Non-working population. The working population means participation in economically productive activity either physical or mental in nature. Thus is involves not only actual work but also supervision and direction; whereas non-working population means, those who did not work at all during the preceeding year (Census of India, 1981).

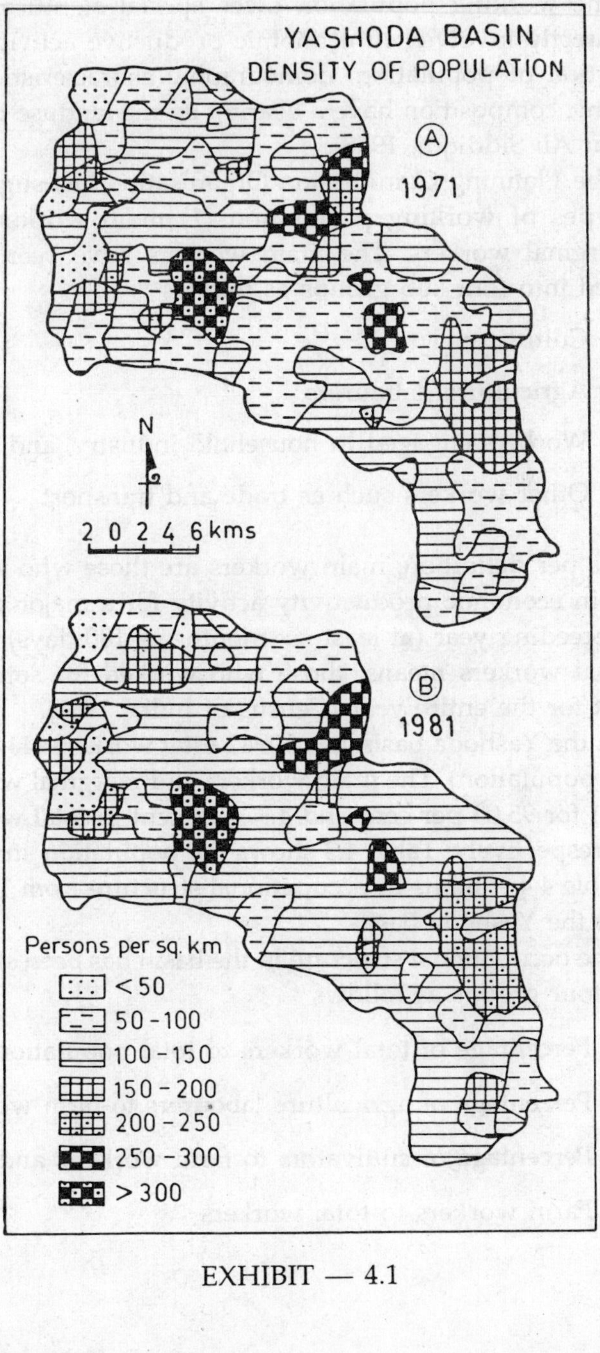
YASHODA BASIN
DENSITY OF POPULATION
A
1971
N
2 0 2 4 6 kms
B
1981
Persons per sq. km
<50
50-100
100-150
150-200
200-250
250-300
>300

EXHIBIT — 4.1

The working population owes special significance as it is directly involved in economic productive activity. The proportion of population, demographic characteristics and economic composition have a bearing on the landuse pattern (Forasat Ali Siddiqui, 1988).

The Planning Commission Organisation has suggested two types of working population (*1*) main workers and (*2*) marginal workers. The main workers have been again grouped into four sub groups as follows :

(*1*) Cultivators

(*2*) Agricultural labourers

(*3*) Workers engaged in household industry, and

(*4*) Other workers such as trade and transport.

As per definition, main workers are those who are engaged in economic productivity activity for a major part of the preceeding year (at least six months or 180 days), while marginal workers means, those who worked for sometime but not for the entire year (Census of India, 1981).

In the Yashoda basin has 50942 total workers (44.56 per cent to population). The main workers and marginal workers account for 95.03 per cent and 4.97 per cent to total workers (1981) respectively. Table 4.3 shows the population structure and Table 4.4 exhibits in occupational structure from 1971 to 1981 in the Yashoda basin.

The occupational structure in the basin has been studied under four groups as follows :

(*1*) Percentage of total workers to total population;

(*2*) Percentage of agriculture labourers to farm workers;

(*3*) Percentage of cultivators to farm workers; and

(*4*) Farm workers to total workers.

Table 4.3
Occupational Structure

Sr. No.	*Categories*	*1971*	*Per cent of total population*	*1981*	*Per cent of total population*
1.	Total population	94741	100.00	114306	100.00
2.	Total workers	39294	41.47	50943	44.58
3.	Cultivators	9903	10.45	14713	12.87
4.	Agricultural labourars	22995	24.27	29868	26.12
5.	Farm workers (Cultivators and Agricultural labours)	32898	34.72	44581	39.00

Source : Author

Table 4.4
The Variations in Occupational Structure (Years 1971 and 1981)

Sr. No.	*Categories*	*Years 1971*	*1981*	*Increase/ Decrease*
1.	Percentage of total workers to total population	41.47	47.44	+ 3.9
2.	Percentage of cultivator to farm workers	30.10	33.00	+ 2.10
3.	Percentage of agricultural labourers to farm workers	69.30	67.00	– 2.90
4.	Percentage of farm workers to total workers	83.72	87.51	+ 3.79

Source : Author

Percentage of Total Workers to Total Population

The spatial distribution of percentage of total workers to total population in the Yashoda basin for 1971 and 1981 are shown in Exhibit 4.2. In 1971, 39294 persons were in the working population category (41.47 per cent to total population) in the basin. Exhibit 4.2/A shows the spatial distribution of percentage of total workers to total population in 1971. It is observed that the total working population increases towards the east and south in the area under study. The central part at Mamdapur (68.75 per cent) is found to have more than 60 per cent workers to total population in the basin and rest has between 20 to 60 per cent workers. Moreover, it is obvious from Exhibit 4.2/A that western part has comparatively less percentage of workers and is involved in agriculture. About 20 to 40 per cent workers to total population was found in seventy villages, whereas, 40 to 60 per cent workers are found in eighty three villages in the Yashoda basin.

The working population in 1981 was 50943 workers (44.50 per cent to total population) in the basin and it is increased by 3.9 per cent after decade (1971 to 1981). Exhibit 4.2/B shows spatial distribution of working population for 1981 in the area under study. It is obvious from the above exhibit that the structure of working population has changed considerably in the study period (1971-1981). The remarkable increase in working population from 20 — 40 per cent to 40 — 60 per cent in the basin after decade. This change is attributed to increasing number of persons in agriculture. Another noteworthy change occurred in the basin in 1981 is that, pockets of 40 to 60 per cent workers are found isolated in nature in central and northwestern parts in the study region. The villages included in this population group are *namely*, Barasonegaon (30.55), Bori (29.22) and Mankapur (37.11) in the northwest part, whereas, Tigaon (57.29) in northern part and Mirapur (59.93), Pathari (32.53), Bhawanpur (53.20), Sonegaon/Khr (58 per cent) in the south in the basin where more per cent of workers are involved in agriculture.

The Percentage of Cultivators to Farm Workers

Exhibit 4.3 gives comparative exposition of spatial distribution of per cent cultivators to farm workers for 1971 and 1981 in the basin. The cultivators include both tenant and owner of farm workers, whereas, farm workers refer to cultivators and agricultural labourers (Census of India, 1981). The involvement of farm workers in agriculture in basin has

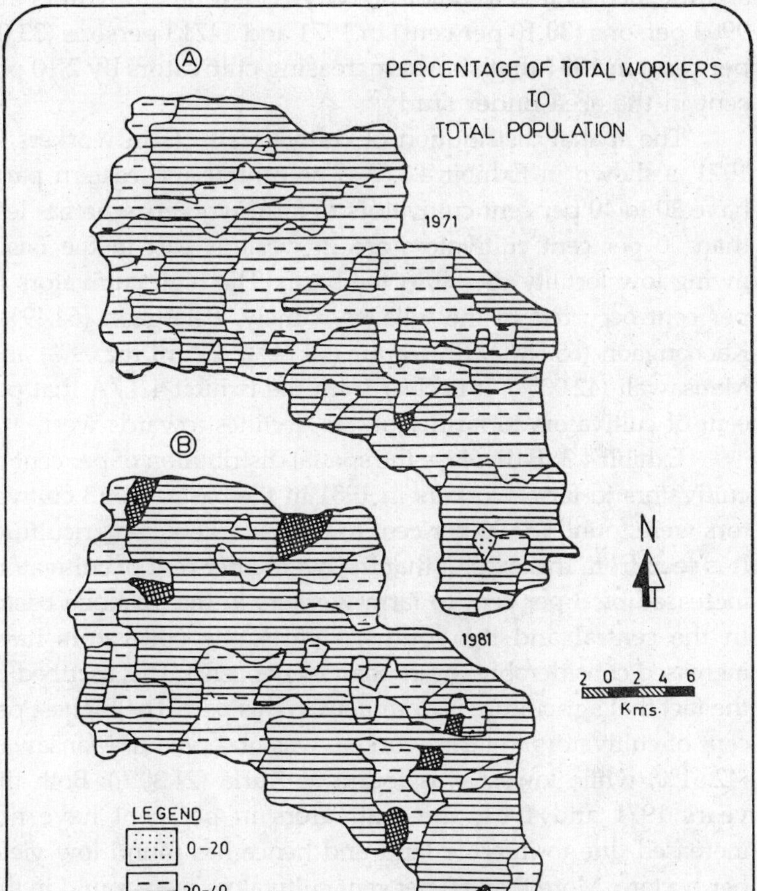
PERCENTAGE OF TOTAL WORKERS
TO
TOTAL POPULATION
A
1971
B
1981
N
2 0 2 4 6
Kms.
LEGEND
0-20
20-40
40-60
> 60

EXHIBIT — 4.2

significance as it is a major activity. The total cultivators are 9903 persons (30.10 per cent) in 1971 and 14713 persons (33.00 per cent) in 1981 that shows increasing cultivators by 2.10 per cent in the area under study.

The spatial distribution of cultivators to farm workers in 1971 is shown in Exhibit 4.3/A. The central and eastern parts have 20 to 40 per cent cultivators to farm workers whereas less than 20 per cent cultivators are in western part in the Basin owing low fertility of soil in the basin. The high cultivators in per cent occurred in the villages namely, Palasgaon (64.49%), Kachangaon (65.58%), Bhiwapur (60.14%), Ekurli (62.09%) and Mansawali (42.00%). It is clear from the Exhibit 4.3./A that per cent of cultivators to farm workers declines towards west.

Exhibit 4.3/B displays the spatial distribution of per cent of cultivators to farm workers in 1981 in the basin. 14713 cultivators were found (33.00 per cent to farm workers) in agriculture. It is seen from this exhibit that the overall per cent of cultivators increased by 3 per cent to farm workers in the Yashoda basin. In the central and the southern parts, the cultivators have increased considerably (more than 60 per cent). This ascribed to the fact that agriculture is dominant in this part. The highest per cent of cultivators to farm workers was observed at Mansawali (42.61%) while lowest was found at Kurla (21.30%). Both the years 1971 and 1981, the cultivators in per cent have not increased due to infertile land and hence has found low yield per hectare. More than 60 per cent cultivators were found in the villages *namely*, Palasgaon (67.25%), Takali (48.40%), Bhiwapur (62.33%), Ajagaon (61.06%), Ekurli (65.11%), Kanchangaon (63.59%), and Mansawali (53.50%) in the Yashoda basin.

The Percentage of Agriculture Labourers to Farm Workers

The per cent of agricultural labourers to farm workers has been worked out at village level for the basin and it was then mapped in Exhibit 4.4 for showing spatial distribution of per cent of agricultural labourers to farm workers for 1971 and 1981. The agriculture labourers in per cent decreased

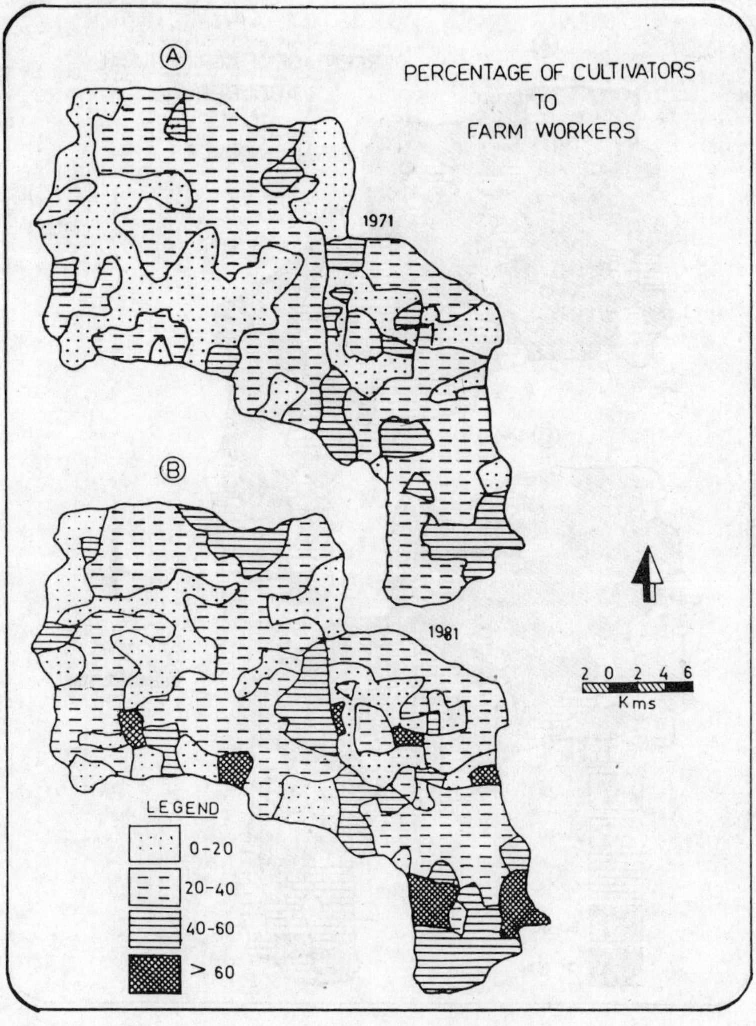
PERCENTAGE OF CULTIVATORS
TO
FARM WORKERS
A
1971
B
1981
2 0 2 4 6
Kms
LEGEND
0-20
20-40
40-60
> 60

EXHIBIT — 4.3

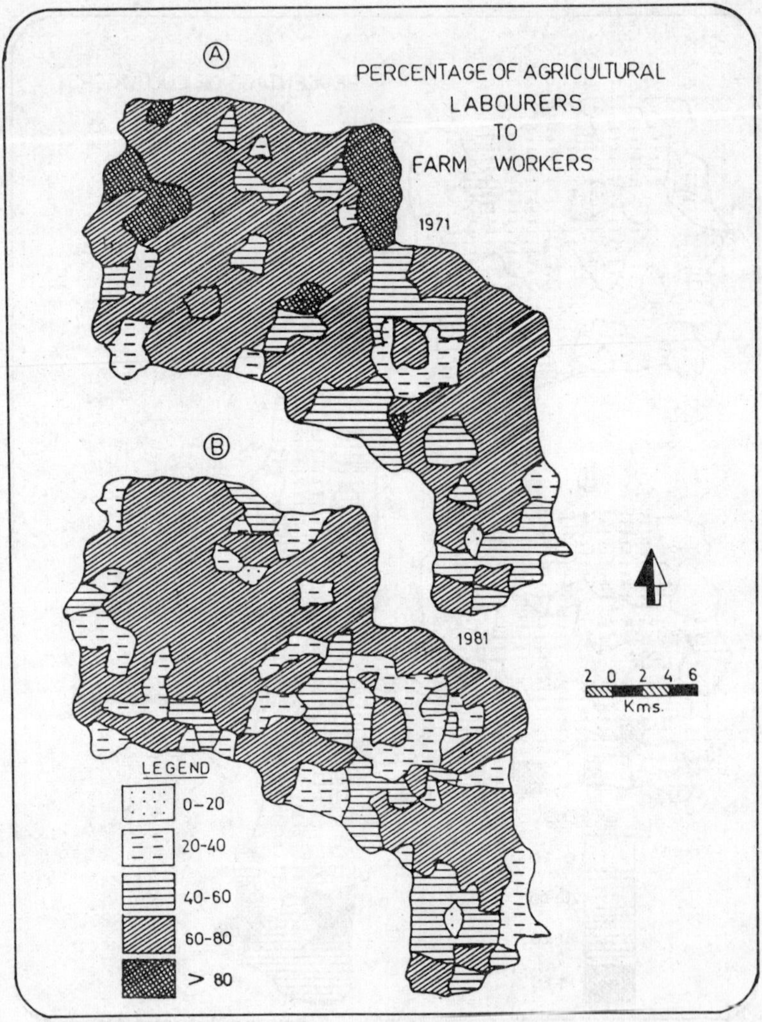
A
PERCENTAGE OF AGRICULTURAL LABOURERS TO FARM WORKERS
1971
B
1981
2 0 2 4 6
Kms.
LEGEND
0-20
20-40
40-60
60-80
> 80

EXHIBIT — 4.4

considerably during study period by 2.90 per cent in the study area. The spatial distribution of agricultural labourers to farm workers is shown in Exhibit 4.4/A for 1971 in the Yashoda basin. It is obvious from this exhibit that the northern part has high per cent of agriculture labourers (more than 60 percent). This high per cent of agriculture labourers were found in the village, *namely*, Sindi/Mgh (82.86), Selukate (64.01), Laxminarayanpur (84.37), Kawatha (63.91), Bodad (82.30), Kelapur (69.09) and Bori (90.69) in the west part and Selsura (91.89), Mamdapur (83.25), Digdoh (84.82), and Mansawali (79.23), in the central part in the basin. Bhagawa has observed less agriculture labourers in per cent (57.14) owing local adverse relief that restricts agriculture practice.

Exhibit 4.4/B displays the spatial distribution of agriculture labourers to farm workers in the Yashoda basin. The main feature of agriculture labourers decreased considerably during study period (2.90 per cent). The Yashoda basin has 29868 agriculture labourers accounting for 67 per cent of the farm workers. It is observed that the proportion of agriculture labourers to farm workers has not changed the nature of spatial distribution pattern than the pattern than that of 1971 for 1981 except more than 80 per cent agriculture labourers category which is concentrated in southern part in the basin. About 60 to 80 per cent agriculture labourers were found to dominate in the central and the southern parts in the study area. (Exhibit 4.4/B). The southern part has concentrated group of agriculture laboursers in the villages, *namely*, Allipur, Takali/Drn, Alamdoh, Rohankheda, Rawani, Pathari, Nimsada and Bhagawa. These villages have not shown any significant increase in the number of agricultural labourers.

The Percentage of Farm Workers to Total Workers

The Yashoda basin is predominantly agricultural in nature. Nearly ninty per cent working force is engaged in agriculture activity. The spatial distribution of farm workers to total workers is shown Exhibit 4.5 for 1971 and 1981. The farm workers include the cultivators and agriculture labourers, whereas, workers refer to participation in economically productive activity (mental and physical) as well as supervision and direct work.

Exhibit 4.5/A shows the spatial distribution of farm workers to total workers for 1971 in the area under study. The region has 32898 farm workers (83.72 per cent to total workers). The per cent of farm workers to total workers has been worked out at village level and it was then mapped and is shown in Exhibit 4.5/A. It is observed from this exhibit that the majority of farm workers are concentrated in central and the southern parts (more than 80 per cent to total workers) in the Yashoda basin in agriculture practice. About 20 to 40 per cent farm workers were found in villages, *namely*, Nagazari, Bid-nagazari, Mankapur, Husenpur, and Renukapur in northeast part of the basin. In the south the villages having 20 to 40 per cent farm workers in the village, *namely*, Loni, Timapur, Dhonapur and Akola. The per cent of farm workers between 20 to 40 in the basin has patchy distribution in the eastern part covering six villages *namely*, Jaitapur, Asala, Fattepur, Kurzadi, Chunhala and Taharpar. Laxminarayanpur, Kawatha, and Chitki have 20 to 60 per cent farm workers in the Yashoda basin.

The spatial distribution of farm workers to total workers in percentage is shown in Exhibit 4.5/B for Yashoda basin in 1981. The basin has 44581 farm workers (87.56 per cent of total workers). The per cent of farm workers has increased by 3.79 per cent in the basin. More than 80 per cent farm workers to total workers has found in many pockets of the basain and these pockets are dispersed in central and southern parts in the study area. The fourteen villages have more than 80 per cent farm workers in central part consisting the villages, *namely*, Dewali, Digdoh, Chikti, Bodad, Jaintapur, Sarur, Pawani, Chanki, Pathari, Takali/Jjr, Selukate, Wadhdha, Chinchala and Takali (Chinj) in the basin. The Yashoda basin has 20 to 40 per cent farm workers to total workers in the north and northwest parts (Exhibit 4.5/B) due to infertile soil

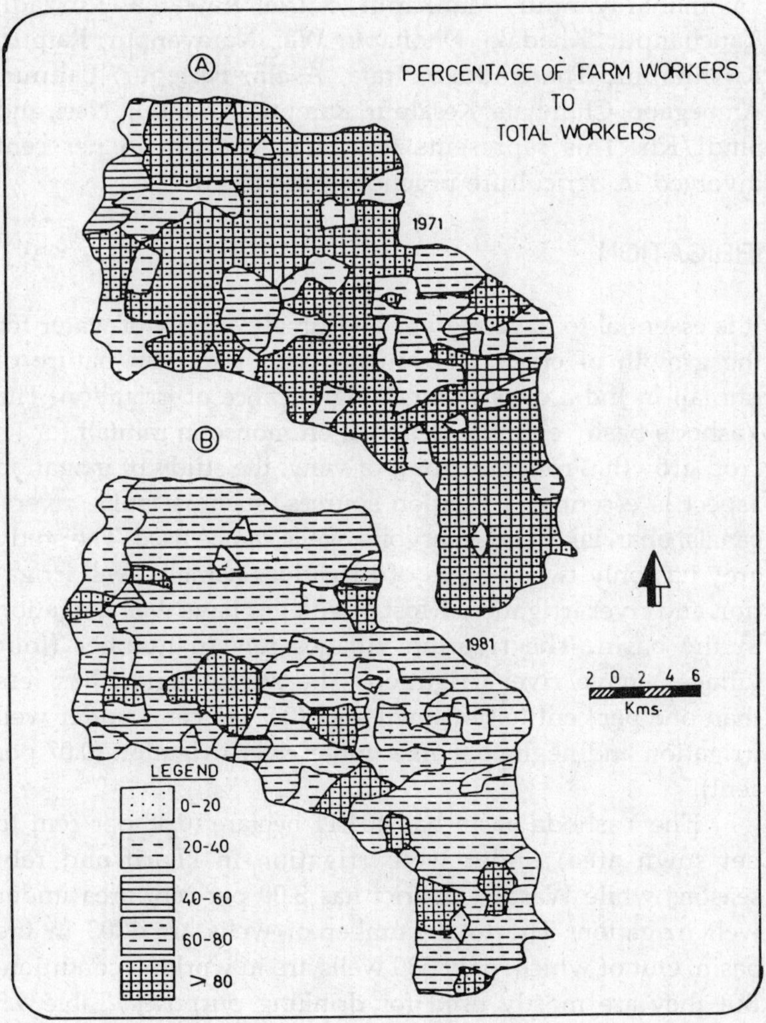
A
PERCENTAGE OF FARM WORKERS
TO
TOTAL WORKERS
1971
B
1981
2 0 2 4 6
Kms.
LEGEND
0-20
20-40
40-60
60-80
> 80

EXHIBIT — 4.5

in the villages, *namely*, Nagazari, Bid-nagazari, Husenpur, Laximanarayanpur, Mankapur, Kurla, Kawatha, Kurzadi, Kanchanpur, Khadaki, Dhonapur Wai, Narayanpur, Raipur, Krishnapur, Thangaddeo, Itala, Asala, Fattepur, Banpur, Ambegaon, Chunhala, Kesalpur, Rampur, Aminpur, Neri, and Sindi/Kh. This represents less farm workers in per cent involved in agriculture practice in the Basin.

IRRIGATION

It is essential to have regular and timely supply of water for the growth of crops. The uneven and uncertain nature of rainfall in India does enhance significance of irrigation. The Yashoda basin, entirely, depends on monsoon rainfall for its crop-growth. From this point of view, the study of irrigation aspect is essential. Irrigation sources, such as tanks, rivers, canals, ohar, lake, and reservoirs, wells can be used. The study arey has only two sources of irrigation, namely, well irrigation and river irrigation. Most of villages have well irrigation in the basin. The riverine villages few in number (four villages) have river-irrigation. The study region has less than one per cent to net sown area (0.5 per cent) under well irrigation and negligible area under river irrigation (0.07 per cent).

The Yashoda basin has 404.11 hectare (0.58 per cent to net sown area) under well irrigation in kharif and rabi seasons while Wardha district has 3.00 per cent area under well irrigation. The total number of wells are 1705 in the basin. Out of which, only 530 wells are in working condition. But they are mostly used for drinking purposes. Table 4.5 displays the availability of well in the basin.

The Exhibit 4.6 displays the spatial distribution of percentage of irrigated area to net sown area in the Yashoda basin. It is noticed from this exhibit that the central and eastern parts have relatively more areal extent under irrigated crops than that of north and the southwest part in the basin. The maximum area under irrigation is found at Dewali (29.3 per cent to net sown area) sand minimum at Mankapur (2.10 per cent to net sown area) in north part in the region. Moreover, it is observed that the land less than 5 per cent to net sown area under irrigated crops show extensive coverage in one hundred and twentynine village (83.1 per cent to total villages) and spread out within the basin (Exhibit 4.6). This fact explains little significance of irrigation in crop cultivation. Four villages occupy more than 15 per cent irrigated area to net sown area where gram, wheat, rice and little vegetables are raised for local consumption. These villages are, *namely*, Kurzadi (27.20 %), Dewali (29.37 %), Talegoan (16.87 %) and Jawalgaon (18.02 %).

Table 4.5
Availability of Wells in the Yashoda Basin

Sr. No.	*Range of wells*	*Number of villages*	*Percentage of villages*
1.	1 - 5	52	33.54
2.	5 - 10	25	16.12
3.	10 - 15	26	16.77
4.	15 - 20	07	04.51
5.	20 - 25	11	07.69
6.	> 725	34	21.93
	Total	155	100.00

Source : Land Records.

The land between 10 to 15 per cent to net sown area under irrigation is noticed in three villages, *namely*, Waigaon (13.10 %), Isapur (10.9 %), and Salod (14.10 %) in the Yashoda basin. (Exhibit 4.6). The farmers are raising crops *namely*, wheat and vegetables in the field. Isapur produces little vegetable for selling it to Dewali market. Nineteen villages appear to have 5 to 10 per cent irrigated area, occupying areal extent in eight patches (Exhibit 4.6). The largest patch in areal extent is confined to east part extending it from Umari to Sonegaon in the basin. This patch covers villages of Umari (6.9 %),

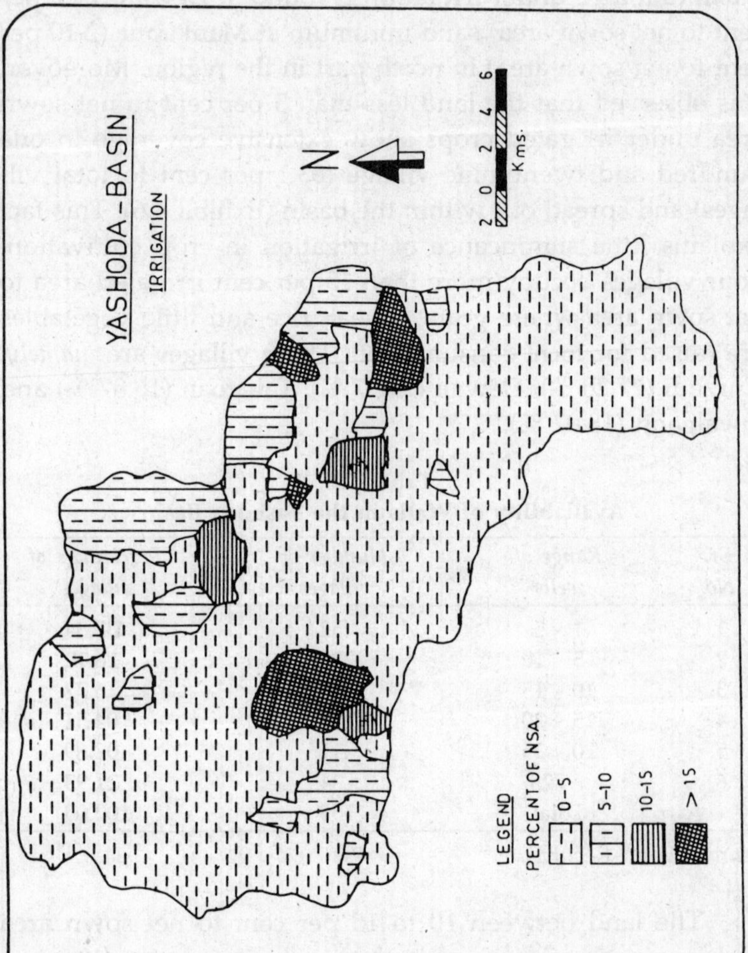
YASHODA-BASIN
IRRIGATION
N
0 2 4 6
Kms.
LEGEND
PERCENT OF NSA
0–5
5–10
10–15
> 15

EXHIBIT — 4.6

Nagthana (6.9 %), Sawangi (6.9 %), Sindi/Mgh (9.1 %), Selukate (7.3 %), Ashratpur (8.8 %), Ashta (5.1 %), and Sonegaon (5.1 %). This little land under irrigation owe proximity to Wardha market which creates the vegetable demand.

MARKETING

The study of marketing is imperative in agriculture geography. it extends selling and purchasing facilities to agricultural commodities. The numerous activities in agriculture seed purchasing and selling goods needs market facility. Multifunction/multidinal function is imperative in modern commercial economy as it reflects the surplus and deficit of total agriculture products. Here, an attempt has been made to examine the spatial distribution of market in the Yashoda basin.

Generally, there are two types of market, one is weekly and other is regulated market. The basin has wide distribution of weekly markets and one regulated market at Dewali has opened up on experimental basis for operation.

The Distribution of Weekly Markets and Regulated Markets

The Exhibit 4.7 shows the locations of weekly markets in the Yashoda basin and it spread out throughout the region. The weekly markets perform socio-economic function. The main function that is served by weekly market is to distribute goods to consumers in the surrounding villages. The marketing in the basin started in medieval period (Wardha District Gazetteer). There are seventeen villages holding such markets nearly five to eight kilometers area. The villages performing markets are shown in Exhibit 4.7. The day of weekly markets and name of villages are given in Table 4.6. The sequence of market day among the villages is best adjusted so as to reduce the competition in the surrounding area in the basin (Table 4.6). The markets in the basin are generally held where village population was between 500 to 2500. It is obvious from Table

4.6 that Wednesday is fixed for six villages namely, Dhamangaon, Kelapur, Dahegaon (Mskn), Aloda, Waddha and Nagazari (Exhibit 4.7). These villages are best adjusted with having 10 to 20 kilometers distance from each market. Four villages have performed Thursday market day in the basin, *namely*, Waifad, Salod (Hpr), Waigaon and Adegaon. Tuesday is fixed to three villages, *namely*, Loni, Allipur and Kanholi in the basin. Tuesday and Friday market held at Giroli, Talegaon and Sirasgaon (Bazar) and Saturday for Palasgaon in the basin. Dewali, previously, a weekly market on Wednesday owing to population growth and their demand has been performed as regulated market. This market centre does not have pronounced impact as it is in primary stage.

Table 4.6
Weekly and Regulated Markets

Sr. No.	*Days*	*Number of villages*	*Name of villages*
(A)	*Weekly market*		
1.	Tuesday	3	Loni, Allipur and Kanholi
2.	Wednesday	6	Dhamangaon, Kelapur Dahegaon (Mskn), Aloda, Waddha, Nagazari,
3.	Thursday	4	Waifad, Salod, Waigaon, Adegaon,
4.	Friday	3	Talegaon, Giroli Sirasgaon (Bazar)
5.	Saturday	1	Palasgaon
(B)	*Regulated market*		
1.	Day	1	Dewali

Source : Author

TRANSPORTATION

The role of means of transport in the economic development is significant in agricultural regions. It acts as a main vehicle for bringing different raw materials, seeds, fertilizers, implements, and distribute the products within the region. The improvement in transport network extends the hinterland of markets and brings isolated productions area into light (Datye,

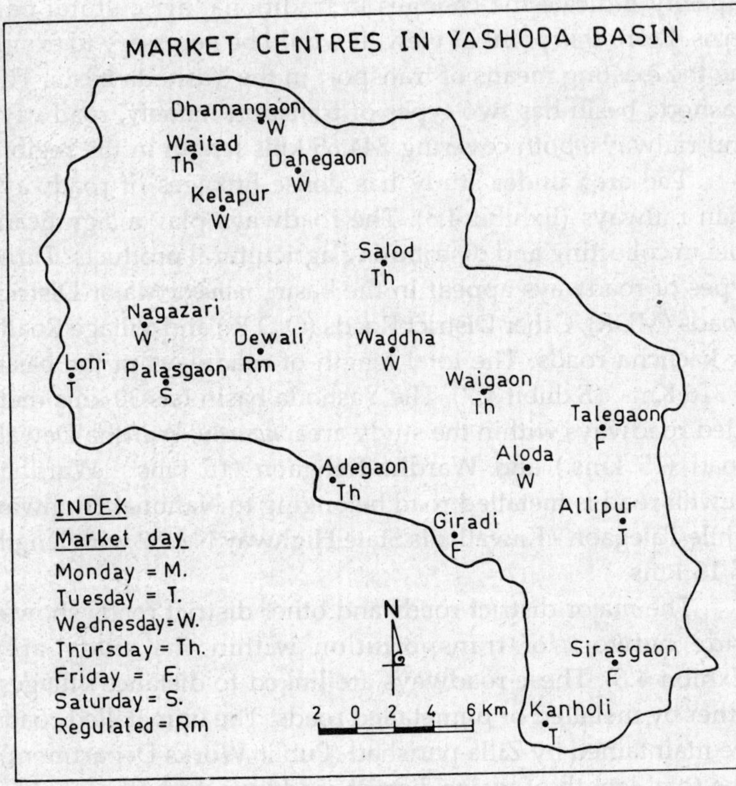
MARKET CENTRES IN YASHODA BASIN
Dhamangaon W
Waitad Th
Dahegaon W
Kelapur W
Salod Th
Nagazari W
Dewali Rm
Waddha W
Loni T
Palasgaon S
Waigaon Th
Talegaon F
Aloda W
Adegaon Th
Allipur T
Giradi F
Sirasgaon F
Kanholi T
INDEX
Market day
Monday = M.
Tuesday = T.
Wednesday = W.
Thursday = Th.
Friday = F.
Saturday = S.
Regulated = Rm
N
2 0 2 4 6 Km

EXHIBIT — 4.7

1983). Such improvement and change in frequency, speed and capacity indicate the changes in traditional agricultural patterns. From this point of view, it would be necessary to examine the existing means of transport in the Yashoda basin. The Yashoda basin has two types of transport namely, roadways and railways, both covering 344.65 kms length in the region.

The area under study has dense linkages of roadways than railways (Exhibit 4.8). The roadways play a significant role in collecting and distributing agricultural products. Three types of roadways appear in the basin, *namely* Major District Roads (MDR), Other District Roads (O.D.R.) and Village Roads or kachcha roads. The total length of roadways in the basin is 316 Kms. (Exhibit 4.7). The Yashoda basin has 30 kms metalled roadways within the study area, *namely*, Wardha Dewali Road (15 kms.) and Wardha-Kangaon (15 kms.). Wardha-Dewali road is metalled road belonging to National Highway while Talegaon - Kawatha is State Highway No. 82 with length of 15 kms.

The major district roads and other district roads show a wide network of transportation within the study area (Exhibit 4.8). These roadways are linked to distance villages, either by metalled or unmetalled roads. The unmetalled roads are maintained by Zilla parishad (Public Works Department). The total length of major district roadways and other district roadway are 42.70 kms. and 41.08 kms respectively in the basin. The village roadways are kachcha in nature and are linked to M.D.R. and O.D.R. These kachcha roads are unsuited for transportation in the monsoon season and it becomes dirty and muddy and hence it reduces the degree of transportation. The most common and traditional modes of transport is bullock cart in the villages to collect and distribute the agricultural products. Trucks are occasionally used for transportation.

It is observed that most villages are linked by bus services of minimum two frequencies to each village. This bus service is owned by State Transport Service, Maharashtra

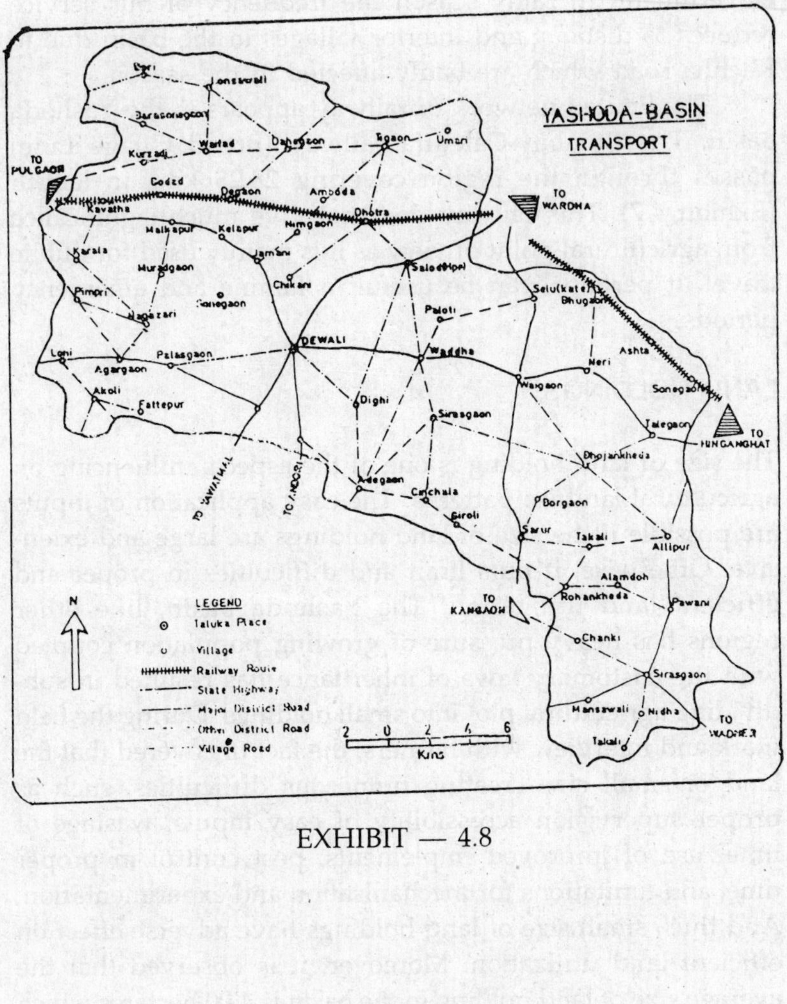
YASHODA-BASIN
TRANSPORT
Warfad
Umari
Kurzadi
TO PULGAON
Godad
Degaon
Amboda
TO WARDHA
Kavatha
Malkapur
Kelapur
Dhotra
Karala
Jamani
Muradgaon
Pimpri
Sonegaon
Chikani
Selu(kate)
Bhugaon
Nagazari
Paloti
DEWALI
Loni
Palasgaon
Waddha
Ashta
Agargaon
Neri
Waigaon
Akoli
Fattepur
Dighi
Sonegaon
Sirasgaon
Talegaon
TO HINGANGHAT
Bhojankheda
TO YAWATMAL
TO ANDORI
Adegaon
Chichala
Borgaon
Giroli
Sarur
Takali
Allipur
Alamdoh
Rohankheda
TO KANGAON
Pawani
Chanki
Sirasgaon
Mansawali
Nidha
TO WADNER
Takali
N
LEGEND
Taluka Place
Village
Railway Route
State Highway
Major District Road
Other District Road
Village Road
2 0 2 4 6
Kms

EXHIBIT — 4.8

Government. In rainy season the frequency of bus service reduces to distanct and interior villages in the basin due to Kachha road which are badly affected in the season.

The limited network of railway appears in the Yashoda basin. The Bombay-Calcutta railway (Broad Guage Line) passes through the region covering 28.25 kms in length (Exhibit 4.7). This railway does not give much significance from agricultural point of view as it is mainly used for public travel. It perhaps, can be useful in famine and emergency periods.

LAND HOLDINGS

The size of land holding is one of the aspects influencing on agricultural landuse patterns. The easy application of inputs are possible if the size of land holdings are large and extensive. Otherwise, it puts limit and difficulties in proper and efficient land utilization. The Yashoda basin, like other regions has heavy pressure of growing population coupled with the customary laws of inheritance has resulted in subdividing agricultural plot into small holdings. During the field work and interview with farmers, the fact uncovered that the land of small size creating numerous difficulties, such as proper supervision accessibility of easy inputs, wastage of time, use of improved implements, pest control in proper time, and limitations for mechanisation and experimentation. And thus, small size of land holdings have adverse effect on efficient land utilization. Moreover, it is observed that the average size of land holding in the basin is 4.00 hectares which is comparatively low than that of Wardha district (7.10) and Maharashtra (4.20). The unequal and uneconomic small size land holdings are observed in western part of the basin resulted backward agriculture. It is, therefore, needed to consolidate the land for efficient landuse in entire basin.

Table 4.7
Route length of Wardha District and Yashoda Basin

(length in kms)

Sr. No.	*Types*	*Wardha District*	*Yashoda Basin*
	Roads :		
1.	National highway	108.00	—
2.	State highway	324.03	55.50
3.	Major District roads	240.51	42.76
4.	Other District roads	168.50	41.08
5.	Village roads	847.81	176.96
	Total :	1688.85	316.30
	Railway :		
(i)	Broad Guage	140	28.35
(ii)	Meter Guage	Nil	Nil
(iii)	Narrow Guage	34	Nil
	Total :	174	28.35

Source : Socio-Economic Abstract, Wardha District, 1985.

AGRICULTURAL IMPLEMENTS

The timely availability of agricultural implements and its proper use in the process of agricultural operation is imperative for proper and efficient landuse in any region. The Yashoda basin, like other regions apply numerous agricultural implements in ploughing, harrowing, levelling, ridging, hoeing and intercultivation. The common implements used in agriculture operation in the basin are iron plough, wooden plough, harrow, Bakhar, Kolpa, Kurhad, Pawada, Khurpi, Vila (Sickle), and Koyata. The types of implements used in agricultural operation indicates the type of farming and resulted level of agriculture in the region. It is observed that the small land holding farmers generally lack required implements which are borrowed from neighbours for time being, while large land holding farmers own themselves most of the agricultural implements and has found improved village due to timely available implements for application in the field.

RESUME

The discussion on socio-economic aspects in this chapter has brought out the nature and major features of various aspects, *namely*, population, occupational structure, irrigation, transportation and marketing in the Yashoda basin. The basin, like other area, has witnessed a steady growth of population with its upward trend. In the decade, 1971 to 1981 the percentage of growth rate had been 17.99 per cent. The average density of population is 132.77 persons per sq. km. in the region (1981). From spatial point of view, the population density increases towards east. Of the total population 87.57 per cent working force is engaged in agricultural pursuits. Of the total working force accounts for cultivators is 28.88 per cent and agricultural labourers is 58.63 per cent. There is negligible area is under irrigation, amounting to 0.58 per cent to net sown area in the basin. Seventeen rural market centres in the basin serve as channels for sale and distribution of agricultural products mainly. The total network of Kachha roads has been estimated to be adequate for circulation of goods and men. Bus services in remote villages are seasonal.

5

General Landuse Pattern

INTRODUCTION

After having uncovered the role of physical and cultural variables in existing landuse of the Yashoda basin in Chapters 3 and 4, a study of landuse classification follows logically in the light of above variables in the area under study. The man has fundamental utility of land in satisfying the human need of food habitation and housing materials. In order to meet the needs, it is essential to choose proper mode of landuse planning and allocation to various ingredients of optimum landuse. Kellong (1980) has rightly pointed out that this calls for a clear understanding of land classification for successful planning and development. The application of various inputs in land may change the allocation of land to different uses. Of all, the factors the conservation of the quality of our socio-economic environment is the most fundamental for the proper use of our land. This statement is true not only of large urban centres but as well for most of the remote areas. The growing pressure of population coupled with an increasing variety of demand on land resources have brought extra pressure on available resources. In order to deal with these and to plan for optimum utilization of land, it is necessary to have accurate and upto date information in all possible details on landuse. It is therefore, the study of classification of landuse pattern in the Yashoda basin would be of great help for preparation of the relative development plan for the basin.

The objective of this chapter is to assess spatial as well as temporal landuse and suggest possible ways and means to improve existing landuse in the basin in the light of physico-socio-economic conditions.

The area of forest cover, net sown area (NSA) area not available for cultivation (ANC) and cultivable waste (CW) have been converted into percentage to total geographical area. Further, these have been used for showing the spatial distribution of land classification with suitable cartographic maps (Exhibits 5.2/A, 5.2/B, 4.2/C and 5.2/D). A line graph (Exhibit 5.1) exhibits the temporal variation of land classification for a period of ten years (1975-1976 to 1985-1986) in the basin. The description of each land classification has been supplemented by numerous spot-inquiries, besides information embodied in the relative District Census Handbook, District Gazetteer and District Socio-Economic Review of Wardha District.

LAND CLASSIFICATION

The land classification aims at dividing land into different categories according to a single factor or set of factors. Therefore, land classification may be of many types depending upon the factors taken into consideration. The land classification has a direct bearing on climatic factors, soil characteristics, slope of land, degree of erosion, water supply, drainage and similar environmental conditions. The landuse capabilities classification portrays physical capability of land to produce over a long period of time for selected uses which can be provide land operation with a basis for actual practice of land (Stamp, 1968).

PREVIOUS LITERATURE

In the recent years several attempts have been made in different countries of world to classify landuses from different point of views and for different purposes, employing varieties of methods. Stamp (1960) has cogently remarked that it is not surprising that the divergence points of view on classification of land prevail. He is well regarded as pioneer in the field of land classification. He has by his example initiated further studies in this direction. In his work entitled "The Land of Britain : Its Use and Misuse" he classified land into six categories, *namely*, (1) Forest and woodland (2) Arable land, (3) Meadow land and permanent grass, (4) Health and moorland, (5) Gardens orchards, nurseries and (6) Unproductive land : such as buildings, mines, wasteland etc.

In the international landuse classification nine major landuse classes have been recognised : (1) Settlement and non-agricultural land, (2) Horticulture, (3) Trees and permanent crops (4) Cropland, (5) Improved permanent pastures (6) Improved grazing land (7) Woodland, (8) Swamps and marshes and (9) Unproductive land.

On the basis of land capability land has been classified in the United States of America. Three classes are determined which are based on the local relief, climate, soil and vegetation. Accordingly, the land falls classified into three categories (1) Land capable of carrying various crops or grass in rotation or permanent (2) Land suitable for the poorer grades of permanent pastrures and (3) Land unsuitable for economic uses as arable, grazing, or aforestation. The soil conservation service and the agricultural landuse planning activities of the U.S.A. Department of Agriculture has classified the land into four capability classes. The National Resource Planning Board of land Classification in U.S.A. classified the land into five categories (1) Land classifications in term of inherent characteristics, (2) Land classification in terms of present day use, (3) Land classification in term of landuse capabilities, (4) Land classification in terms of recommended uses and (5) Land classification in terms of programme implementation.

In U.S.S.R. the following types of land are classified as under (1) Land belonging to the urban rural localities to the industrial and transport department, (2) Arable lands and perennial plantation, (3) Natural grass lands, (4) Forests, (5) Bogs and peats bogs and (6) Unsuitable land.

A.K. Phibrick (1980) has classified land in China into following categories : (1) Agricultural land 60 per cent or more under cultivation, (2) Agricultural land at least 20 per cent under cultivation, (3) Advancing agriculture and aforestation, (4) Upland with some cultivation and patches of forests, (5) Natural forest, (6) Steppe and deserst grazing land, some cultivation and patches of forests and (7) Major drought prone areas (1960-61).

In Japan the land utilization is classified into rice land, high land, with dry agriculture, forested land and uncultivated land. The Landuse survey in Japan has been conducted since 1953 and upto the present day a large number of its sheets have been published. On the basis of these landuse maps (eleven) main categories of land have been recognised which have further been sub-divided into many micro-subdivision. The categories are (1) Paddy area, (2) Upland field, (3) Tree crops, (4) Forest area, (5) Grass land, (6) Other indicated by salt bed, barren land, fish culture area and boundaries of different fields, (7) City and town, (8) Transportation, (9) Land improvement and land conservation facilities, (10) Special facilities, such as, power plants, mine, transportation substations and (11) Land features, such as, swampy lands and peat-bog.

In India various schemes have been proposed to classify the land into different uses. The National Atlas Organisation in 1957 classified the land into nine categories (1) Forest, (2) Scrub, (3) Arable land with trees, (4) Plantation, (5) Pasture, (6) Wasteland, (7) Alpine grass and scrub and (8) Glaciated region. The Domodar Valley region has classified land into ten major categories : (1) Field crops, (2) Orchards, (3) Dense forests, (4) Light Forests, (5) Non-agricultural land, (6) Unproductive land, (7) Water bodies, (8) Culturable waste, (9) Villages and (10) City and towns.

Prior to the year 1951 the landuse statistics collected by the Ministry of Agriculture, Government of India was arranged according to the following classification : (1) Total geographical area, (2) Area under forest, (3) Area not available for cultivation, (4) Current fallow land, (5) Other uncultivated land and (6) Net area sown.

In 1954, E. Ahmad classified the rural land into following categories with detailed coloured technique for landuse mapping : (1) Tripple cropped land, (2) Double cropped land, (3) Single cropped land, (4) Fallow land, (5) Land under grove and orchards, (6) Land under scrubs and grasses, (7) Land permanently under water, (8) Built-up land, (9) Land and communication, (10) Barren land and (11) Forests.

Landuse records department has officially classified land under following categories : (1) Reporting area for land utilization purposes, (2) Forest, (3) Barren and unculturable land, (4) Land put to non-agricultural uses (i) culturable waste, (ii) Permanent pastures and other grazing land, (5) land under miscellaneous tree crops and groves not included in net area sown; (i) Current fallows (ii) Other fallow land, (6) Net sown area, (7) Area sown more than once and (8) total cropped area.

It would be convenient for a clear understanding to condense the above mentioned twelve categories into five categories as (1) Forest land, (2) Net Sown area, (3) Land not available for cultivation, (4) cultivable waste and (5) Fallow land.

TEMPORAL VARIATIONS IN LANDUSE

The temporal variations in landuse pattern in the Yashoda basin have been studied for a period of ten years (1975-76 to 1985-86) and possible causes of changing landuse are interpreted. The investigator could not succeed in uncovering temporal variations of landuse for consecutive years due to paucity of data for the years concerned. However, alternative year has been taken into consideration for showing temporal variations in landuse pattern in the area under study.

The temporal variations in landuse for Yashoda basin are studied in five categories as follows :

1. Net Area Sown (NSA)
2. Cultivable Waste (CW)
3. Fallow land (FL)
4. Area not Available for cultivation, (ANC), and
5. Forest/Forest Cover (FC)

The changes occurred during the period of study are interpreted as follows

Net Sown Area (NSA)

The net sown area is steadily increasing since 1975-76 to 1985-86 (Exhibit 5.1). It is seen from Exhibit 5.1 that 49.39 per cent area was under cultivation in 1975-76 and it has been stepped to 79.58 per cent area under cultivation in 1985-86, registering an increase by 30.18 per cent. This increase may be attributed to increasing awareness of improved agricultural techniques and introduction of the use of new improved seeds by farmers. From 1975-76 to 1977-78, net sown area increased by 8.93 in 1977-78 and 1979-80, it has increased by 7.15 per cent, from 1979-80 to 1981-82, there is a slight increase (1.59 per cent). It increased 8.45 per cent from 1981-82 to 1983-84 and finally rate of increase is low during the period of 1983-84 to 1985-86 (4.03 per cent). There is an average 3.00 per cent increase except for the year 1979-80 to 1981-82 (only 1.59 per cent). This significant increase in net sown area may be due to more land under trees and shrubs, brought subsequently under cultivation. Therefore, other types of land have continuously declined from 1975-76 to 1985-86 (Exhibit 5.1 and Table 5.1). Other types of land previously considered cultivable waste is being used by the farmers for cultivation

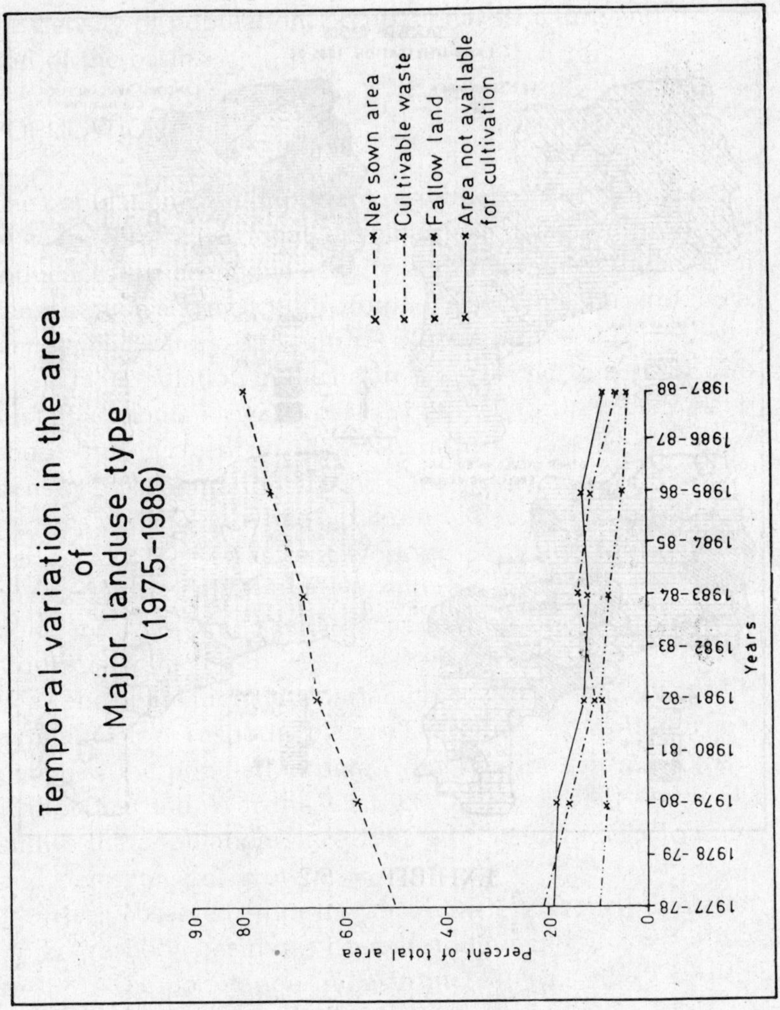
Temporal variation in the area
of
Major landuse type
(1975-1986)
Net sown area
Cultivable waste
Fallow land
Area not available
for cultivation
Percent of total area
90
80
60
40
20
0
1977-78
1978-79
1979-80
1980-81
1981-82
1982-83
1983-84
1984-85
1985-86
1986-87
1987-88
Years

EXHIBIT — 5.1

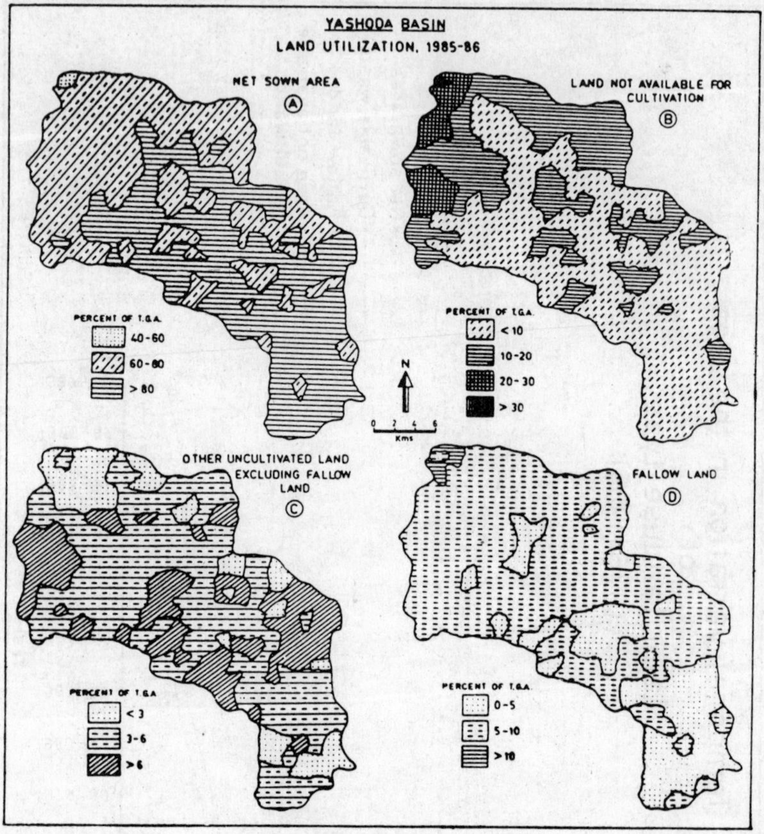
YASHODA BASIN
LAND UTILIZATION, 1985-86
NET SOWN AREA
A
PERCENT OF T.G.A.
40-60
60-80
> 80
LAND NOT AVAILABLE FOR CULTIVATION
B
PERCENT OF T.G.A.
< 10
10-20
20-30
> 30
N
0 2 4 6
Kms
OTHER UNCULTIVATED LAND EXCLUDING FALLOW LAND
C
PERCENT OF T.G.A.
< 3
3-6
> 6
FALLOW LAND
D
PERCENT OF T.G.A.
0-5
5-10
> 10

EXHIBIT — 5.2

for growing crops. Moreover, increasing awareness, and mounting pressure of population on landuse pattern has brought this land under cultivation.

Land Not Available for Cultivation

This category includes the land put to non-agricultural uses, barren and uncultivated land. The area under this category has declined steadily from 1975-76 to 1985-86 in the study region. The total decline during the study period is 11.31 per cent (Table 5.1). There is slight decline during the last decade. The land not available for cultivation has been decreasing due to the increase in the land under cultivation. (Exhibit 5.1 and Table 5.1). The Exhibit 5.1 reveals the temporal variations in land out of non-agricultural uses and barren and uncultivated land. Non-agricultural land has been substantially decreased for the period 1975-76 to 1985-86 (7.65 per cent) than barren and uncultivated land (3.76 per cent). More land that in the past has been put to non-agricultural use, brought under cultivation due to feasibility of such land for cultivation in the basin whereas barren and uncultivated land could not be brought under cultivation being stiff and hard soils. Therefore, only 3.76 per cent barren and uncultivated land has been brought for raising additional crops.

Cultivable Waste

In the Yashoda basin, cultivable waste indicates steady decline during the period of ten years (Exhibit 5.1 and Table 5.1). The cultivable waste includes such sub-types as permanent pasture and other grazing land, miscellaneous tree crops and groves not included in net sown area and culturable waste.

The trend of decline in the cultivable waste is shown in Exhibit 5.1). The total decline is 12.96 per cent from 1975-76 to 1985-86 (Table 5.1) that shows downward trend. The permanent pasture and other grazing land declined by 3.78 per cent, miscellaneous trees and groves by 5.44 per cent and culturable waste declined by 4.74 per cent. There was a large decline in miscellaneous tree and groves due to notable awareness increase among farmers.

Table 5.1
Temporal Variations in General Landuse Pattern
(1975 to 1986)

(Area in Percentage)

Sr. No.	Landuse types	Years 1975-76	1977-78	1979-80	1981-82	1983-84	1985-86
1.	Net sown area	49.39	58.32	65.51	67.10	75.55	79.58
2.	Land not available for cultivation	20.13	18.18	11.80	13.18	11.51	8.82
(a)	Land put to non-agricultural uses	12.88	12.00	6.30	9.15	7.00	5.13
(b)	barren and uncultivated land	7.45	6.18	5.50	4.03	4.51	3.69
3.	Cultivable waste	20.03	15.50	11.42	11.54	10.73	7.07
(a)	Permanent pastures and other grazing land	9.11	8.00	7.19	6.27	6.23	5.33
(b)	Miscellaneous trees crops and groves not included in net sown area	5.52	2.50	2.13	3.37	2.35	0.08
(c)	Cultivable waste	8.40	5.00	2.10	2.00	2.15	1.88
4.	Fallow Land	10.45	8.00	11.27	8.08	4.21	4.49
(a)	Current Fallow	7.15	6.00	7.27	5.00	2.21	1.87
(b)	Fallow land other than current fallow	3.30	2.00	4.00	3.08	2.00	2.62
5.	Forest	Nil	Nil	Nil	Nil	Nil	Nil

Source : Village Revenue Records for 1975 to 1986.

The decline trend in case of permanent pasture and other grazing land during study period is 12.98 per cent. (Table 5.1). The lowest decline of this land found from 1983-84 is 0.04 per cent. The area under miscellaneous tree crops and groves has been decreasing in the Yashoda basin and near about 6.00 per cent land has brought under cultivation during the study period.

Fallow Land

The fallow land includes current fallow and land other than fallow. The current fallow means land kept uncultivated for regaining fertility of soil and other purposes during the agricultural year. Other fallow land means land kept uncultivated more than five years due to various reasons *i.e.* non-availability of capital, lack of agricultural know-how. In study region both current fallow and other fallow land show decline trend during the period of ten years 6.28 per cent and 0.68 per cent respectively (Table 5.1). This fact suggests that less land under other fallow has been brought under cultivation. Moreover, there is a fluctuation in the area under fallow land from 1975-76 to 1985-86.

Forest

Either tracts of reserved forest or protected forest do not occur in the Yashoda basin. Only trees and isolated shrubs are limited in number in western part in the villages, namely, Pimpri, and Nagzari. The open trees have been cut down by local farmers to cultvate land. Hence the study region is almost devoid of forest-cover. Around the source region of river Yashoda, the land is treeless and barren.

COMPARATIVE STUDY OF LANDUSE IN YASHODA BASIN, WARDHA DISTRICT AND MAHARASHTRA

Table 5.2 shows the area under different landuse categories in the Yashoda basin, Wardha district and Maharashtra. Net sown area in the basin is relatively more under cultivation (79.58 per cent) than Wardha District (68.92 per cent) and Maharashtra (59.49 per cent). The less hectare net sown area in Wardha district and Maharashtra attributed to highly diversified relief whereas the Yashoda basin is marked out its plain surface with black cotton soil throughout its length and breadth. Moreover, absence of forest cover within the basin does not restrict land for cultivation is also responsible for the entension of land under crops. Table 5.2 shows that negligible land under miscellaneous tree crops and groves (0.09 per cent) in the Yashoda basin. It is less than Wardha district (1.93 per cent) and Maharashtra (0.60 per cent). The area under other types of land, current fallow, cultivable waste, barren and uncultivated land have remained relatively unchanged in the basin (Table 5.2).

SPATIAL VARIATIONS IN LANDUSE

The spatial pattern in landuse in the Yashoda basin is the result of interaction between physical environment and socio-economic environment. But the impact of regional and local factors is clearly evident from the landuse patterns. Besides these factors, amount of rainfall exerts profound influence on the types of landuse in the Yashoda basin. The overall landuse has been categorised into different subtypes on the basis of recommendations made by Food and Agricultural Department, Government of India. These subtypes are as follows :

(1) Forest

(2) Land not available for cultivation :

  (*a*) Land put to non-agricultural uses,

  (*b*) Barren and uncultivable land

(3) Other uncultivated lands excluding fallow land :

  (*a*) Permanent pastures and other grazing land.

  (*b*) Land under miscellaneous tree crops and groves not included in net area sown.

Table 5.2

The Comparative Landuse of Yashoda Basin, Wardha District and Maharashtra

Sr. No.	*Categories*	*Yashoda Basin*		*Wardha District*		*Maharashtra*	
1.	Forest	Nil		66400	(10.55)	530900	(17.26)
2.	Landput to non-agricultural uses	4424.92	(5.13)	34900	(5.54)	993000	(3.22)
3.	Barren and uncultivated land	3176.66	(3.69)	8800	(1.39)	1733000	(5.63)
4.	Permanent pastures and other grazing land	4597.09	(5.34)	47800	(7.60)	1591000	(5.17)
5.	Miscellaneous trees crops and groves not included in net sown area	77.47	(0.09)	12200	(1.93)	186000	(0.60)
6.	Culturable waste	1420.45	(1.64)	8800	(1.39)	993000	(3.22)
7.	Current fallow land	1611.85	(1.87)	9700	(1.54)	852000	(2.77)
8.	Fallow land other than-current fallow	2262.13	(2.62)	111000	(1.76)	802000	(2.60)
9.	Net Sown area	68517.43	(79.58)	420900	(66.92)	18299000	(59.49)
	Total	86088.00	(100.00)	628900	(100.00)	30758000	(100.00)

Source : Compiled by Author.

(4) Fallow Land :

(a) Current fallow land and,

(b) Fallow land other than current fallow, and,

(5) Net area Sown.

The above mentioned landuse categories have been used to examine the spatial pattern of landuse in the Yashoda basin. The per cent of each landuse type has been calculated to total geographical area of the basin. The area and percentage of landuse types has given in Table 5.3.

Table 5.3
General Landuse of Yashoda Basin (1985-1986)

Sr. No.	*Landuse types*	*Area (In Hectare)*	*Percent of Total Geographical Area*
1.	Net Sown Area	68517.43	79.58
2.	Land not available for cultivation	7601.58	8.83
3.	Area under cultivable waste	8095.01	7.07
4.	Area under fallow land	3873.97	4.50
5.	Forest	Nil	Nil
	Total	86088.00	100.00

Source : Village Revenue Records, 1985-86.

Net Sown Area

The net sown area includes land actually under food, cash and fodder crops. The Yashoda Basin has 79.58 per cent land under cultivation to total geographical area showing relatively more area than that of Wardha district (66.92 per cent) and Maharashtra (59.49 per cent). This peculiarity of the land pattern of the basin can be accounted for by its low-level plain and total absence of forest cover.

Exhibit 5.2/A reveals the spatial variations in the per-cent of area of net sown area to total area in the Yashoda basin. It is observed from this exhibit that net sown area in-creases from west to east and from north to south in the basin. This is the outstanding trend of net sown area. This can be explained by the fact that northwest part of the Yashoda basin occupied by Tigaon Hill that is almost barren, unsuitable and low fertility soil. As a result, comparatively less percentage of area has been brought under cultivation. Remaining part of the Yashoda basin (central, eastern and southern parts) is low-lying plain with medium to deep black soils, suitable for extensive land under plough.

The maximum per cent of land under net sown area is found at Chanki (91.15 per cent) in south part of the Yashoda basin, whereas minimum net sown area is found at Mankapur (40.00 per cent), in the source region of the river in hilly tract. The deep black soil at Chanki in lower Yashoda basin has been enabled to bring maximum land under crop.

Fiftynine villages have observed more than 80.00 per cent net sown area in the central and eastern parts of the basin (Exhibit 5.2/A). This region is a low level plain of fertile tract of soil except few villages where local relief does not permit to increase net sown area. The extensive fertile tract in the central and eastern part appears to be conducive for raising various crops on black soil and hence attempts are being made to bring every piece of land under agriculture. In the deep black soil zone, net sown area more than 85 per cent found in the villages, *namely*, Sirasgaon (Bazar), Chanki, Mansawali, Pathari, Takali (Drn), Allipur (85.00), Alamdoh (85.10), Gadegaon (85.10), Pawani (86.92), and Kachangaon (90.00). These villages lie in the lower Yashoda basin where local physical and cultural environment permits to cultivate more land under plough.

The category consisting 60 to 80 percent net sown area to total area is sown in Exhibit 5.2/A and is found in the western and the northern parts except small isolated extreme northwest chunk in the Yashoda basin. Some similar isolated pockets are found in the central and eastern parts. In the western and northern parts of the Yashoda basin have a continuous patches with 60 to 80 per cent land under crops.

Eightly five villages have 60 to 80 per cent net sown area in the basin. The minimum net sown area is found in Husenpur (61.58) in northwest part while maximum net sown area is found in Waifad (80.00) in the Yashoda basin. There are thirteen small pockets of 60 to 80 per cent net sown area in the basin distributed in central and eastern parts in the study region. The villages having more than 70 per cent net sown area in the basin are *namely*, Padhegaon (76.33), Muradgaon (63.85), Dagapur (72.32), Thangadheo (72.30), Narayanpur (70.00), Ajanavati (71.52), Banpur (65.32), Ashartpur (65.91), Fattepur (71.82), Kurzadi (73.50), Chunhala (75.12), Babapur (64.60), Bhugaon (77.50), Jawalgaon (71.32), Ashta (76.00), Itala (75.31), Dewangan (69.30), Ambapur (70.00), Mamdapur (76.35), Mund (72.52), Babpur (73.58), Mendukdoh (76.37), Gadegaon (Rt.) (71.62), Bhagawa (70.20), Sonpeth (77.78), Ajanavati (74.58), Bopapur (73.58) and Ekurli (75.22).

About 40 to 60 per cent net sown area is found in the extreme northwest part in the Yashoda basin. This portion covers three villages, Bidnagarari (51), Mankapur (40), and Nagazari (43). This region is the source of Yashoda river where hilly tract dominates the landscape. The hilly nature of the land and rugged local topography act as constraint on the expansion of net sown area (Exhibit 5.2/A).

Land Not Available for Cultivation

This is sub-divided into following types :

(*a*) Land put to non-agricultural uses: This type includes land occupied by settlements, roads, railways streams, canals and rivers.

(*b*) Barren and uncultivable land includes outcrops of hills and mountains. The small part of this land can be brought under cultivation of very high costs. Generally, barren and uncultivated land is associated with poor soils, heavy rainfall and intense erosion.

In the Yashoda basin, the land not available for cultivation accounts for 8.83 per cent (7061 hectares) to total geographical area. This type shows considerable variations within the basin ranges from 4.21 per cent to 29.20 per cent under land not available for cultivation. This land is marked by gradual decrease from the west to the east. The maximum land under this category is confined to Nagazari (29.20 per cent) in the west, whereas the minimum land not available for cultivation is found at Borgaon (4.21 per cent) in the east part of the Yashoda basin (Exhibit 5.2/B).

The less than 10 per cent of land is not available for cultivation is in the central and the eastern parts of the basin (Exhibit 5.2/B). The maximum per cent under this category occurs in the central part of the basin at Agargaon (10 per cent) and the minimum per cent is noted at Borgaon (9.20 per cent) in the eastern part of the basin. Seventy three villages fall in this class.

Seventyone villages fall in the range of 10 to 20 per cent of land not available for cultivation in the study area. This region is marked out in the north, west, and the central parts in the basin. Most of these villages are jaxtaposed to the Tigaon Hill or its offshoots in the northern part of the basin, accounts for by its rugged topography 10 to 20 per cent of land not available for cultivation are noted at Padhegaon (14.03), Ashartpur (4.09), Narayanpur (14.79), Meghapur (20), Malegaon (15.30), Babapur (20), Ajanavati (17.16), Thangaddeo (17.11), Dagapur (10.50), Narayanpur (11.20), Bopapur (12.35), Raipur (11.80), Krishnapur (10.75), Banpur (16.68), Fattepur (12), Chunhula (11) and Kesalpur (16.20).

The maximum land not available for cultivation is found at Babapur (20) whereas the minimum at Dhamangaon (10.40) (Exhibit 5.2/B).

Twenty to 30 per cent of the total areal extent of land not available for cultivation lies in the western part of the basin (Exhibit 5.2/B). The land under this category is marked by gradual increase towards eastern part in the study region. In all the ten villages have 20 to 30 per cent of land under this class. This maximum land under not available for cultivation lies at Nagazari (29.20 per cent), near source region of the basin and the minimum at Kurzadi (21.03 per cent). Out of these ten villages, seven are located in northwest part in the study area (Exhibit 5.2/B). These villages are *namely* Nagazari (29.20), Mankapur (27.85), Husenpur (24.21), Renukapur (22.30), Panchanpur (21.68), Bori (24.62), and Bara-sonegaon (22.50). This tract is hilly, not suitable for cultivation by the prevelant techniques of agricultural practices.

The land not available for cultivation is slightly more (8.83 per cent) than the corresponding in Wardha district (6.93) and came close to that of the Maharashtra (8.85 per cent).

Fallow Land

The Yashoda basin has 3873.98 hectares of its total area under fallow land (4.50 per cent). The fallow land is generally, divided into two categories : (*a*) current fallow land, and (*b*) other than fallow.

The current fallow land includes the land which is not cultivated during the current year due to a variety of reasons *i.e.* as phase of rotation, for regaining fertility or due to some other constraints. The land than current land includes arable area which is taken up for cultivation but has gone temporarily out of cultivation for a period of not more than five years.

The magnitude of fallow land varies from village to village in the basin. It ranges from 1 per cent to 13 per cent fallow land gradually diminishes towards east (Exhibit 5.3/C). It is observed that maximum land under fallow category is confined to Renukapur (12.54 per cent) in the west part whereas the minimum per cent fallow land at Chanki (1.89 per cent) in the south part in the basin. The fallow land is spread out over north, west and central parts covering 113 villages.

Less than 5 per cent of fallow land lies in the south and central parts with isolated patches. Thirtyfive villages have less than 5 per cent land under this category in the basin

(Exhibit 5.2/C). The maximum fallow land is found at Dighi (4.03 per cent) and minimum at Chanki (1.89 per cent). The decline of fallow land to the south can be attributed to the marked fertility of local soils.

5 to 10 per cent of fallow land is mainly concentrated in the northern, central and western parts in the basin (Exhibit 5.2/C). The maximum fallow land is visible at Mankapur (9.89 per cent) in northwest and the minimum is at Bhugaon (5.12 per cent) in the study region. The general trend of declining fallow land towards east due to existence of fertile soil raising for cultivation. Five village (3.22 per cent of the villages) in the basin have more than 10 per cent fallow land confining portion in the northwest part where it ranges from 10 to 13 per cent of fallow land. These villages are *namely*, Nagazari (11 to 21), Renukapur (12.54), Kanchanpur (10.79), Bori (10.15) and Husenpur (12.03). The extension of the cultivated area has not been attempted at present due to the local diversified relief features of the surface.

Cultivable Waste

The cultivable waste land includes other uncultivated lands excluding fallow land. This category is divided into three types. (*a*) permanent pastures and other grazing lands, (*b*) miscellaneous tree crops and groves and (*c*) culturable waste.

The permanent pastures and grazing lands include all land, are under grass-cover, government and private land; or permanent pastures which are kept reserved as a village common grazing ground or vast tract of protected land, not open for free grazing and unreserved grass lands. The miscellaneous tree crops and groves include land under grasses, bamboo, bushes, and other groves for fuels etc. which are not included under orchards or forests are included in this category and, the land not cultivated during the preceeding five years is called culturable waste.

The Yashoda basin has 7.07 per cent of total areal extent (6095.01 hectares) under fallow land (Table 5.2). It is slightly less than that of Wardha District (10.92 per cent) and Maharashtra (9 per cent). The cultivable waste is observed in varied ranges within the basin. One comes across the maximum cultivable waste at Muradgaon (9.84 per cent) in west and the minimum at Bara-sonegaon (1.93 per cent) in the northwest part in the study region. The extensive spread of cultivable waste land exceeding 6 per cent lies in the western, central and northern parts in the region. Four major pockets of cultivable wasteland are visible in the basin. One pocket is in the west, and other one is in east. Other two pockets have in the central part in the Yashoda basin. The pocket consisting of eight villages in the west of the basin. The eastern pocket consists of thirteen villages and other two pockets comprised of seventeen villages. Third and fourth pockets of cultivable waste spread out in a scattered manner. The maximum cultivable waste land is confined to Muradgaon (9.84 per cent), and minimum is met at Sindi (Maghe) (6.07 per cent) in the Yashoda basin. The central pocket consisting villages, *namely*, Chitaki (7.13), Sekapur (8.11), Kurla (6.7), Nandpur (8.33), Muradgaon (98.4), Pimpri (6.19), Khadaki (9.27), Nagazari (6.46). The eastern pocket comprising villages, *namely*, Wallapur (7.43), Jawalgaon (6.91), Ashta (7.30), Sonegaon (Stn.) (8.09), Mirapur (6.95), Dattapur (8.43), Sultanpur (7.39), Talegaon (6.82), Ajanavati (8.69), Sonpeth (9.30), Bhojankheda (6.97), Babapur (8.44), and Dapuri (Exhibit 5.2/C).

The central part has two pockets having more than 6 per cent of cultivable waste encompassing seventeen villages (10.96 per cent) to total villages *namely*, Ekpal (6.31) Ajanavati (6.17), Thangaddeo (7.43), Dagapur (9.28), Narayanpur (8.66), Ashratpur (9.18), Meghapur (6.19), Narayanpur (7.50), Babapur (8.44), Chikhali (7.86), Bopapur (6.49), and Raipur (8.04), Dighi, (7.30), Krishnapur (8.10), Sirasgaon (8.25), Wai (7.30), and Chinchala (8.75).

The cultivable waste with 3 to 6 per cent of areal extent to total is observed in central part is patchy in nature. The expansion of cultivable waste has not been attempted at present due to fertile soil largely used for raising crops. The northwestern part in the basin has observed cultivable waste less than 3 per cent of total areal extent in two villages, *namely*, Waifad (2.92 per cent) and Barasonegaon (1.92 per cent) measures as minimum cultivable waste in the Yashoda basin. The barren and local unfavourable relief features of the surface does not spare piece as cultivable waste.

RESUME

The disucssion in the previous pages has been able to uncover the realistic picture of temporal and spatial distributional patterns of landuse in the basin. Both temporal and spatial landuse in the basin are the outcome of physio-socio-economic conditions prevalent in the study region. The Yashoda basin has 79.58 per cent net sown area to total geographical area. This shows higher hecterage under plough in the basin than that of Wardha District (66.92 per cent) and Mahrashtra (59.49 per cent). The existing net sown area has attained the present extent after 1975 with the introduction of high-yield variety of cotton and hybrid jowar. The central and eastern parts of the basin have higher concentration of net sown area on medium to deep black soils tract. The maximum per cent of net sown area is found at Chanki (91.15 per cent) and lowest at Mankapur (40 per cent). The net sown area increases towards the east from west in the basin. Other landuse types have less hectarage kept uncultivated in 1985-86. The land not available for cultivation is 8.33 per cent, cultivable waste 7.07 per cent and fallow land is 4.50 per cent. Forest is conspicuous by its absence with only scanty and sparse trees in the basin. Therefore, vegetation cover has no impact on general landuse in the Yashoda Basin.

In order to understand the temporal aspect of general landuse types, the investigator has covered the time-span of 1975-76 to 1985-86. It is observed that the net sown are

(Exhibit 5.1) after 1977-78 records a visible extension with the interest of farmers to bring every piece of land under plough. The total net sown area increased by 30.90 per cent during the study period, whereas, other categories of landuse types have been steadily reduced land hectarage. The land not available for cultivation is reduced by 2.31 per cent, cultivable waste by 12.9 per cent and fallow land is by 5.98 per cent to the total geographical area. The decline in the later categories in the basin due to the increasing demand of food on one side and increasing awareness among farmer to agricultural practice.

6

Agricultural Landuse Pattern

INTRODUCTION

After unfolding the various categories of general landuse in the Yashoda basin in the earlier chapter, it would be worthwhile to explain agricultural landuse pattern in the study area for individual crops to explicate the crop growth and crop pattern. The term, "Agricultural Landuse" denotes the extent of the gross cropped area during the agricultural year under various crops. It is the result of the decision made by the farmers regarding the choice of crops and methods for production. Thus, this decision-making is based on not only the physical constraints and limitations but also on farmers perception of the total environment. His perception of environment is related to contents and nature of available informations, much of which is based on traditional approach. The physical as well as cultural environment reflect on crop growth and production.

The present chapter attempts to expound the agricultural landuse, its growth and crop pattern in relation to the physio-socio-economic conditions prevalent in the basin and highlights the croplanduse in the context of existing crop ecology, their spatial distribution, methods of farm operation and to discuss the possible causes for the existing croplanduse pattern in the basin.

The areas under various crops have been obtained for 155 villages from concerned talukas and it was converted into percentage to net sown area, which are, later on, used for depicting the distribution of individual crops. The distribution patterns show variations in agricultural landuse in the Yashoda basin. The information collected during field work sustain explanations. The District Census Handbook, Wardha District Gazetteer and Socio-economic Abstract of Wardha have been further used for explaining the agricultural landuse patterns. An analytical exposition on crop patterns have been attempted.

AGRICULTURAL LANDUSE PATTERNS

The spatial and temporal variations in agricultural landuse patterns have been studied in the area under review. The spatial distribution patterns were studied for single year *i.e.* 1985-86, while temporal study were made for the period of 1975-76 to 1985-86. The investigator could not succeed in displaying temporal variations for consecutive years due to lack of data for the taluka concerned. Hence, every alternative year has been taken into account for temporal study of agricultural landuse in the Yashoda basin.

CROPS AND SEASONS

There are two agricultural seasons in the Yashoda basin, *viz.* kharif and rabi. The kharif crops are sown in June or July and harvested at the end of October or early November, while rabi crops are sown in October or mid-November and harvested in month of February or March. The Kharif crops include rice (Oryza Sativa Linn), local Jawar (Andropogen Sorghum), Cotton (Gossypium spp.) Mug (Phaselous Radiatus Roxb), Mot (Phaselous aeonitifolius Jaca), hybrid Jowar (Sorghum Bicolor), Tur (Cajancus Indicus), Groundnut (Arachis Hupogaea Linn), Seasamum (Sesamum Indicum

Linn) and vegetable, while rabi crops are Wheat (Tritium Yulgare/Satiumm), Gram (Cicer Arietinum Linn) and Linseed (Linum Usitatissimum Linn).

TEMPORAL VARIATIONS IN AGRICULTURAL LANDUSE PATTERNS

The crop pattern in any region cannot remain static due to the variations in the rainfall amount and nature of inputs and environmental instability. Moreover, introduction of new high yielding varieties of seeds, irrigation facilities, and technical know-how are responsible for temporal changes. The factors for such changes in cropping patterns differ from village to village and region to region. Therefore, it is worthwhile to study isolated causes of change occurring through space and time.

Exhibit 6.1 displays the temporal variations in cropping pattern in the Yashoda Basin from 1975-76 to 1985-86. The identified possible main features of temporal variations in cropping pattern can be summaried as below :

(*1*) Cotton is predominant in the area under review since 1975. The areal extents of cotton has recorded steady increase during the study period. The total increase of cotton is 17.11 per cent from 1975-76 to 1985-86. (Table 6.1). Cotton was cultivated in 21.29 per cent in the year 1975-76 and is culminated to 38.50 per cent onwards except 1983-84 year (Table 6.1). During 1979-80 and 1981-82 the net sown area under cotton reveals highest increase (9.70 per cent). The introduction of new high yielding varieties of seeds *i.e.* H_4, H_8, H_6 and L_{1003} may be responsible for increasing area and yield of cotton in the basin.

(2) Hybrid Jowar occupies second place after cotton in the Yashoda basin. Before 1975-76, farmers had been generally cultivating local Jawar. Local Jowar is known

as "Gawathi Jowar". This Gawathi Jowar was widely cultivated throughout the basin but due to its low yield per hectare, this variety was replaced by high yield species of hybrid jowar by farmers. Table 6.1 shows the temporal variations in hybrid jowar from 1975-76 to 1985-86. during this period the total hectarage increased by 5.63 per cent. In 1975-76 the hybrid jowar was cultivated on limited area (9.37 per cent), later on, it increased to 15.00 per cent in 1985-86.

(3) Local Jowar stands third in areal extent in the basin. The hectarage of to this crop during the study period (1975-76 to 1985-86) has steadily declined. Before the introduction of hybrid jowar (high yield variety), basin had extensive cultivation of local jowar. Table 6.1 clearly indicates that local jowar was widely cultivated in 1975-76 (34.50 per cent to net sown area) in the area under review. But the hectarage under local jowar has reduced further to 20.00 per cent (Table 6.1) during the study period. The maximum hectarage of local jowar declined in 1975-76 by 8.83 per cent to total areal extent of this crop due to preference for hybrid jowar.

(4) The area under cereals, *namely,* mug, mot, udid, kultha have declined in the following years after 1980 due to preference for improved seeds of new specis in order to get high returns per hectare from the fields (such as hybrid jowar and Hn, Ho, H_8, L_{1003} cotton varieties. It is clear from Table 6.1 that hectarage under other cereals was 25.38 per cent to net sown area in the beginning (1975-76) which was further reduced 10.06 per cent in 1985-86. This fact can be explained by the shift of concentration from other cereals to cotton and hybrid jowar etc. The rate of decline under other cereals is 2.0 per cent in the basin. In 1977-78 the land under other cereals was 22.43 per cent. The

Table 6.1
Temporal Variations in Agricultural Landuse (1975-76 to 1985-86)

Crops	*Years*												*Total*
	1975-76		*1977-78*		*1979-80*		*1981-82*		*1983-84*		*1985-86*		
	Area (in %)	*Decrease/ increase*	*Area (in %)*	*Decrease/ increase*	*Area (in %)*	*Decrease/ increase*	*Area (in %)*	*Decrease increase*	*Area (in %)*	*Decrease increase*	*Area (in %)*	*Increase decrease*	*Total*
Cotton	21.39	Nil	23.80	2.41	28.10.	4.30	35.30	7.20	36.25	0.95	38.50	2.25	17.11
Hybrid Jowar	9.37	Nil	11.20	1.83	12.00	0.80	12.80	0.80	13.15	0.45	15.00	1.95	5.67
Local Jowar	34.50	Nil	25.67	-8.83	21.00	-4.57	19.00	-21.10	17.30	-1.50	14.50	-2.8	-20.00
Other Cereals	25.38	Nil	24.43	-2.95	19.40	-3.03	18.99	-0.41	16.86	2.13	15.32	-2.8	-10.04
Total Pulses	12.24	Nil	12.00	-0.24	14.30	2.30	10.60	-4.30	11.37	0.77	9.35	-2.02	-2.89
Total Oilseeds	2.60	Nil	3.53	0.93	3.35	-0.18	2.20	-1.15	4.09	1.89	5.80	1.79	3.28
Other miscellaneous crops	1.12	Nil	1.37	0.25	1.75	0.38	1:11	-0.64	0.98	-0.13	1.45	0.47	0.33

Source : Revenue Records

Note : The Year 1975-76 is considered as base year.

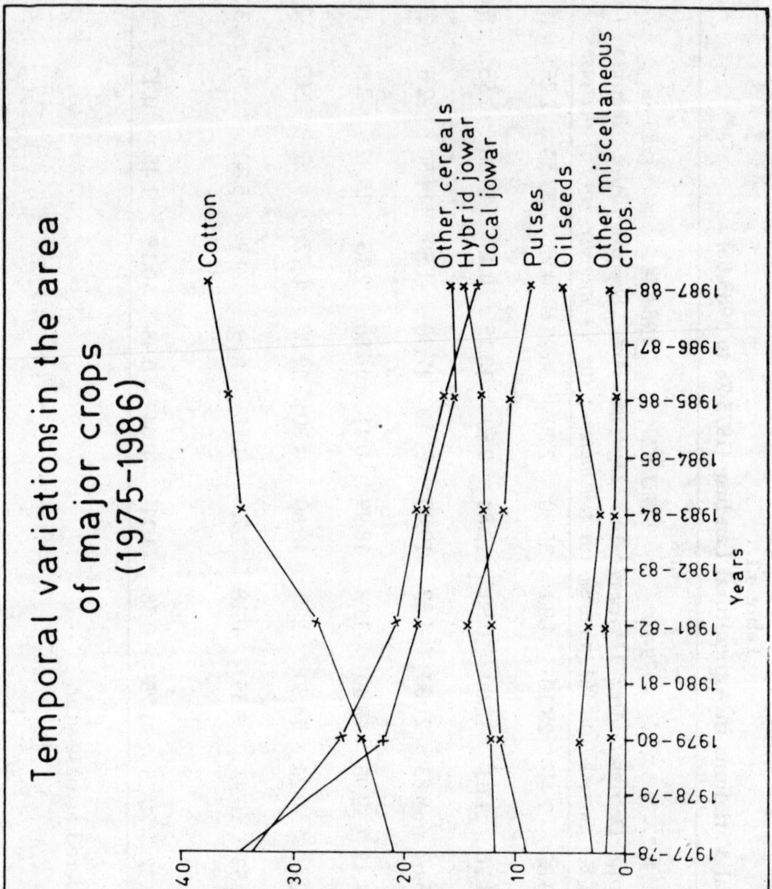
Temporal variations in the area
of major crops
(1975-1986)
Cotton
Other cereals
Hybrid jowar
Local jowar
Pulses
Oil seeds
Other miscellaneous
crops.
Percent of N.S.A.
40
30
20
10
0
1977-78
1978-79
1979-80
1980-81
1981-82
1982-83
1983-84
1984-85
1985-86
1986-87
1987-88
Years

EXHIBIT — 6.1

declining trend under other cereals has continued for further years. It was 19.92 per cent in 1979-80 and 16.80 per cent was for 1981-82 in the Yashoda basin.

(5) The areal extent under pulses in the basin had successfully cultivated before 1975-76 on the area of 12.24 per cent to total net sown area (1975-76). The reduction in area under pulses was registered for the following years. It has been reduced by 9.35 per cent in 1985-86. The total decline of pulses from 1975-76 to 1985-86 is 2.89 per cent except in 1979-80 and 1983-84 where it has 20.30 per cent and 0.77 per cent respectively in the basin.

(6) Table 6.1 shows 3.28 per cent to net sown area under oilseeds has increased during study period (1975-76 to 1985-86). It is noticed from Exhibit 6.1 that the oilseed was found to 2.6 per cent area in 1975-76, while it increased to 5.88 per cent to net sown area in 1985-86. The maximum hectarage under oilseeds was registered in the year 1983-84 (4.09 per cent) in the basin.

(7) Other miscellaneous crops have remained nearly unaffected during the study period except for the year 1975-76 which is increased to 1.75 per cent (Table 6.1). The total decline recorded was about 0.33 per cent. Evidently, this is insignificant change in crops hectareage. It was observed that the area under oil-seeds was 1.12 per cent in 1975-1976 and 1.45 per cent in 1985-86.

The temporal variation reflects major changes with upward trend in the area under cotton and hybrid jowar and hence, the hectarage of these crops have steadily increased due to increasing demand of other crops, *namely*, wheat, gram, groundnut and mot have not registered increase in hectarage in the basin.

SPATIAL ANALYSIS OF AGRICULTURAL LANDUSE

(Crop ecology, land operation and crop distribution)

The Yashoda basin is an essentially agricultural dominant region involving 87.57 per cent working force in agricultural practice. The crops, *namely*, cotton, hybrid jowar, local jowar, wheat, rice, mug, mot, gram, linseed and groundnut are cultivated in the basin. The variations in areal extent under these crops are mostly depending on adoptation of improved varieties of seeds for sowing, local environment and traditional approach of farmers in the area under review. Besides these, soil types and slight rainfall exert influence on the crop cultivation and cropping pattern. Exhibit 6.2 portrays the agricultural calendar for the village in the basin. It shows agricultural active season and relax season in the study area. Moreover, this Exhibit 6.2 demarcates rabi and kharif season prevailing in the basin. It is observed from Exhibit 6.2 that agricultural activities are dominant in kharif season than that of rabi as major crops are raised in rainy period. The relative importance of rabi and kharif crops is shown in Exhibit 6.3/A. It is seen that more than three-fourth area is under plough in kharif (83.58 per cent to net sown area) and less than one fourth is under rabi (16.46 per cent to net sown area) season in the basin. The relative importance of individual crop in kharif and rabi are shown in Exhibit 6.3/B.

The relative significance of crops, and their spatial variations in the area under review has studied in detail with studying crop ecology, land operations and spatial distribution of crops in the Yashoda basin.

Cotton (Gossypium Spp.)

Cotton is the principle crop in the Yashoda basin. It is rainfed crop cultivated in kharif season. This crop is raised in one hundred and fifty villages occupying 26387.95 hectares (36.38 per cent to net sown area). The average annual rainfall amount and soil types in the basin favour the cultivation of cotton in every village.

AGRICULTURAL CALENDER FOR YASHODA BASIN

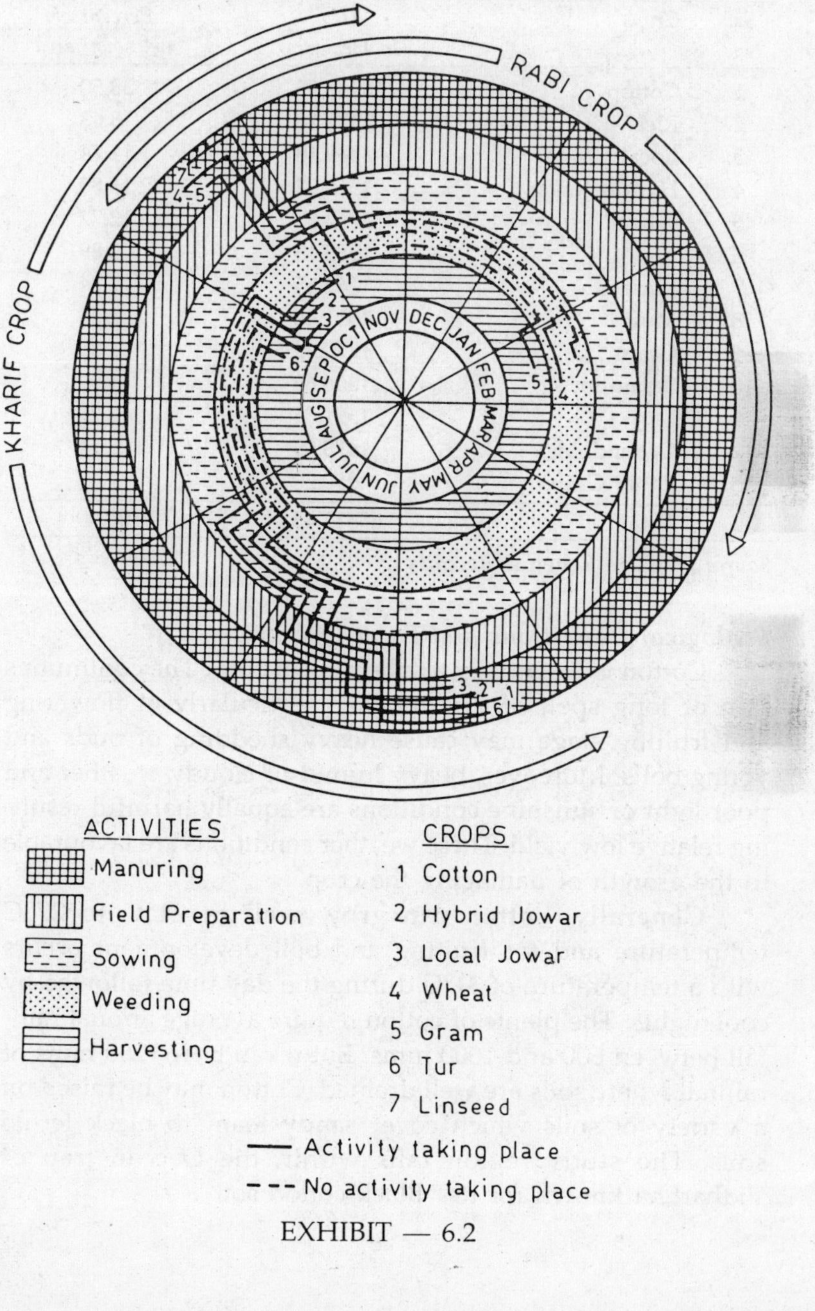

EXHIBIT — 6.2

Table 6.2
Agricultural Landuse of Yashoda Basin (1985-86)

Sr. No.	*Crops*	*Area (In Hectares)*	*Percent of net sown area*
1.	Cotton	26387.95	38.50
2.	Rice	2351.18	3.43
3.	Local Jowar	8049.18	11 74
4.	Hybrid Jowar	11566.92	16.88
5.	Wheat	8754.57	12.77
6.	Tur	1983.13	2.89
7.	Linseed	1262.15	1.84
8.	Mug	1611.46	2.35
9.	Gram	1270.92	1.85
10.	Sesamum	1306.62	1.92
11.	Mot	1551.06	2.28
12.	Groundnut	1473.09	2.14
13.	Other Crops	950.20	1.47
	Total	68517.43	100.00

Source : Village Revenue Records.

Ecological Conditions and Farm Operation

Cotton is essentially a sub-tropical crop. The continuous rain or long spell of dry weather particularly at flowering and fruiting stage may cause heavy shedding of buds and young bolls. Moreover, heavy humid or cloudy weather and poor light or sunshine conditions are equally harmful resulting relative low yield. These weather conditions are favourable to the growth of damaging the crop.

Generally, cotton can grow well in 21°C to 27°C temperature and for fruiting and boll development occurs with a temperature of 31°C during the day time followed by cool nights. The plants of cotton require average annual rainfall between 600 and 1000 mms. But it can bear 1250 mms of rainfall where soils are well drained. Cotton may be raised on a variety of soils which cover sandy loam to black fertile soils. The study region falls within the Deccan trap of Vidharbha known for its "Black cotton soil".

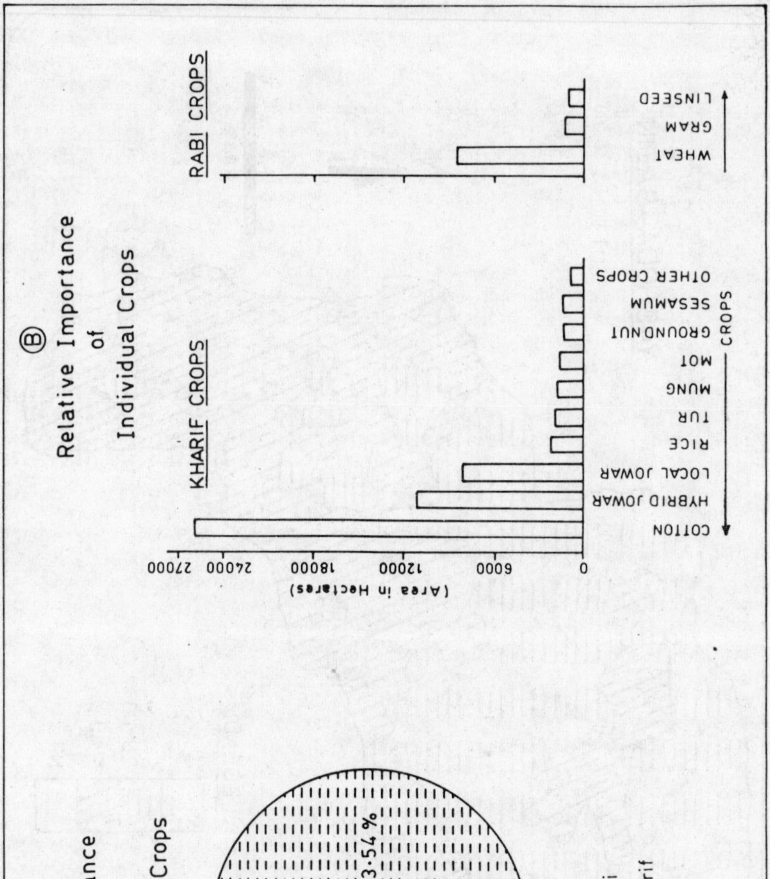

Ⓐ Relative Importance
of
Rabi and Kharif Crops
16·46%
83·54%
Index
Rabi
Kharif
Ⓑ Relative Importance
of
Individual Crops
KHARIF CROPS
RABI CROPS
27000
24000
18000
12000
6000
0
(Area in Hectares)
COTTON
HYBRID JOWAR
LOCAL JOWAR
RICE
TUR
MUNG
MOT
GROUNDNUT
SESAMUM
OTHER CROPS
CROPS
WHEAT
GRAM
LINSEED

EXHIBIT — 6.3

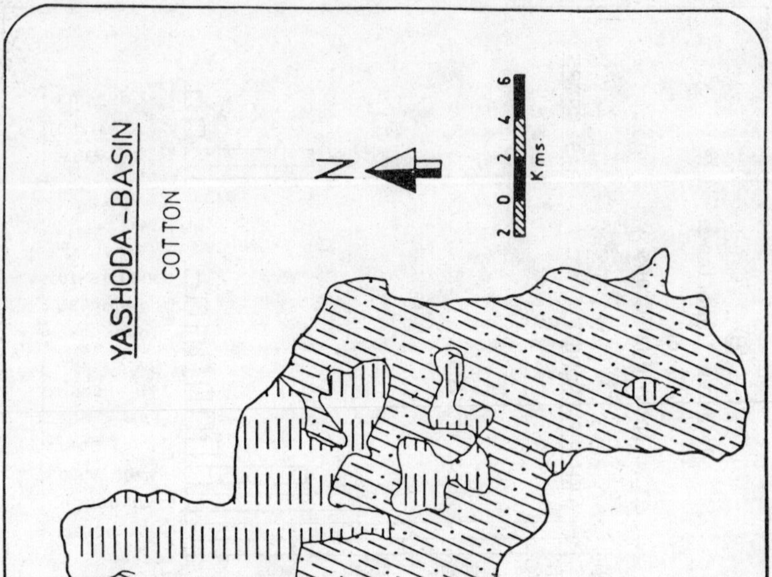

EXHIBIT — 6.4

The farm operations of cotton commences in April and May. The deep ploughing by wooden plough or iron plough is followed by harrowing three to four times. The sowing of cotton, in the basin, is started with the onset of the monsoon in the second week of June. Hand dibbling as well as drilling in rows methods are practiced for cotton sowing. Hand weeding is done with three or four weeks after sowing and is followed by two to three interculturing operation, *namely*, "Dawara", bladed hoe at an interval of 20 days. Organic manuring and chemical fertilizers are added to fields to raise the cotton yield.

Cotton is harvested in three to six pickings as the bolls nature. It means the *Kapus* (seed cotton) is removed from bolls from cotton plants by labourers, particularly, open bolls are collected and picked. This picking period in the basin extends from October to January.

Spatial Distribution

The Exhibit 6.4 reveals the cotton distribution in the Yashoda basin. It is widespread in the region due to favourable geographical conditions. The average annual rainfall amount between 700 and 1100 mms and coarse shallow to deep black soils determine the areal extent under cotton. High yielding varieties of cotton are sown throughout the basin. The regur soil in the basin is the key factor for raising cotton. Farmers, now a days, are adopting improved seeds, chemical fertilizer, and pesticides in cultivating cotton. It is seen from the Exhibit 6.4 that the areal spread of cotton increases from the west to the east and southeast coinciding with suitable soil types and rainfall amount.

The cotton distribution in the basin falls into three categories as under :

(*1*) The area more than 40 per cent to net sown area in the basin between 900 and 1100 mms.

(2) The second category is confined to the land occupying between 20 and 40 per cent to net sown area, and

(3) Less than 20 per cent to net sown area under cotton on coarse shallow soils and the annual rainfall between 800 and 900 mms.

The area of land occupying more than 40 per cent of net sown area under cotton is found in the south and central parts of the basin (Exhibit 6.4). The maximum hectarage of cotton is noticed in the riverine tract in the central part in the Yashoda Basin. The hectarage under cotton increases from the north to the south with increasing soil fertility. Rotha has observed maximum cotton hecterage (41.00 per cent to net sown area) in the northern part in the basin.

First pocket of sixtytwo villages in the basin are found major concentration under cultivation of cotton. The major concentration of the cultivation of this crop is in the south and the east parts in the villages *namely*, Takali/Drn (42.78), Chikhali (41.45), Sonegaon/Bai (40.6), Adegaon (41.69), Chinchala (42.59), Takali/Csnj. (41), Aloda (41.35), Mamdapur (41), Pathari (41.52), Nimsada (41.19), Allipur (10.9), Rohankheda (40.10), Gadegaon (41.15), Pawani (41.13), Chanaki (40.4), Sonegaon/Knr (40.1), Kachangaon (41), Sonegaon/Bai (42.21), Mansawali (42.03), Nidha (41.15), Yarangaon (42), Kanholi (41.1), and Takali/Jjr (41). The tract of the Yashoda basin specialises in new varieties of seeds of high yield.

Another pocket of cotton having more than 40 per cent to net sown area is observed in the Doab of rivers Punsoda and the Bhadali (Exhibit 6.4) comprising seven villages in the basin, *namely*, Bodad (42.8), Yesgaon (43.1), Kelapur (42.45), Degaon/stn. (43), Neri (40.72), Amboda (41.76) and Dahegaon/Mskn (43). Besides these, two pockets under cotton are noted at Fattepur (42.39), Agargaon (41.15), and Rotha (41) in the basin.

The second category of 20 to 40 per cent area under cotton to net sown area occurs in eightysix villages in north and west parts in the basin (Exhibit 6.4). The areal extent under this crop is mainly confined to coarse shallow soils and coarse to medium deep soils where the annual rainfall ranging between 800 and 1000 mms. The cotton hectarage in this category increases towards the east from west. Maximum cotton hectarage recorded at Taharpur (40 per cent to net sown area) and minimum at Bara-Sonegaon (20.20) in north-west part on upland and in fertile region. Other villages with 20 to 40 per cent to net sown area under cotton are *namely*, Babapur (39.00), Ashartpur (39.94), Fattepur (38), Kurzadi (36.97), Chunhala (35), Asala (39), Rampur (38), Neri (39.31), Siwapur (37), Dattapur (34.57), Ambapur (36.18), and Sonpeth (39.29) in the east part on the fertile land where annual rainfall is comparatively higher than that of the west and central parts of the Yashoda basin (more than 900 mms).

In the western and the northern parts of the basin, the villages occupying 20 to 40 per cent area to net sown area under cotton are Laxminarayanpur (30), Kawatha (32.45), Chitki (33), Sekapur, (35.7), Malkapur (3.33), Karla (36), Nandpur (35), Jamani (29.1), Pimpri (36.4), Khadki (33), Muradgaon (37.9) and Sonegaon/Abja (35), while, northern part on coarse shallow soils. Cotton has comparatively less than that of rest part of the basin. The villages having located in this part has low yield per hectare due inferior soils, hence cotton is cultivated between 20 and 40 per cent of net sown area in the villages, *namely*, Lonsawali (30), Dorali (31), Dhamangaon (32.4), Wathoda (33.2), Sindi/Kh. (35), Amala (36), Tigaon (34), Umari (37), Kurzadi (30.6), Waifad (39.4), Dhotra (39) and Nagthana (39.5) in the Yashoda basin (Exhibit 6.4).

Hybrid Jowar (Sorghum bicolour)

The hybrid jowar is a kharif crop in the Yashoda basin. It is grown on 11566.92 hectares (16.88 per cent to net sown area) throughout the region. It is large in size like local jowar hence hybrid jowar is known as *Great Millet*. This is improved variety of seeds than that of local jowar producing high yield per hectare. The introduction of hybrid jowar since 1980 has brought change in scenario in the cropping pattern. Hybrid jowar stands second in rank holding dominant position among food crops in the basin.

Ecological Conditions and Farm Operation

Hybrid jowar can grow successfully in tropical and sub-tropical climate. It requires temperature between 21°C and 27°C and average annual rainfall between 700 and 1100 mms. Hybrid jowar needs higher temperature during its maturity stage. Prolong drought and excessive rainfall are harmful in proper growth of plant. It can cultivate on all soil type, but black soil produces high yield per hectare than that of other soils. Hybrid jowar is grown on medium to deep black soils in the basin. The field of hybrid jowar is prepared by ploughing in April and May followed by two to three harrowing in order to absorb water immediately. While it is sown mixed with mug. Hybrid jowar is generally sown in first week of June in monsoon. Dibbling method is common for sowing in the basin. The timely application of chemical fertilizers and sprayings are essential to assure high yield. Three to four sprayings and three bags of Suphala are given for proper growth after every fifteen days. Interculturing is made in order to control weeds and loosening the soils. The dried hybrid jowar is cut down by labourers with sickle. Threshing is made on threshing machine. Then it is dried and stored safely.

Spatial Distribution

Hybrid jowar is successfully grown in the Yashoda basin except northwestern hilly part (Exhibit 6.5). The improved seeds of this crop assured high yield per hectare. The average annual rainfall between 900 and 1100 mms and prevalent temperature permit for growth of hybrid jowar in the basin. The medium to deep black soils in the basin are also responsible to widespread of this crop and hence the spatial distribution and its areal extent is confined by these factors. The spatial distribution of hybrid jowar is shown in Exhibit 6.5. It is obvious from Exhibit 6.5 that this crop is found in three categories as under :

(1) Highly concentrated area of hybrid jowar on medium black soils is found in thirtynine villages;

(2) Moderately concentrated area under hybrid jowar is observed in one hundred and nine villages dispersed in all soil types; and,

(3) Low concentration of hybrid jowar is found in north-west part confined area on coarse shallow soils in four villages in the basin.

More than 20 per cent area of villages under hybrid jowar is found in two pockets in the central and eastern parts in the area under review. Other eight small pockets are located in the central part being isolated in their nature. The major pockets located in central part is confined to medium black soils in the villages, *namely*, Dewali (27.10), Chikti (21), Padhegaon (21.10), Solad/Hpr (23.10), Jamatha (21.33), Selukate (20.22), Chunhala (23.10), Aminpur (23.30), Taharpur (21.19), Rampur (21.30), Wallapur (23.50), Neri (23.11), Bhiwapur (25), Mirapur (25.30), and Sultanpur (22.72).

Other pocket is found in the villages *namely*, Bori (20.19), Wathoda (20.92), Sindi/Kh. (20.10), Tigaon (20.52), Kawatha (20.51), Agargaon (20.32), Timapur (20.32), Akoli (20.51), Dhonapur (20.31), Narayanpur (20.51), Ambapur (20.10), Aloda (20.19), and Ajagaon (0.35) in the basin.

The second category of the villages where hybrid jowar is found between 10 and 20 per cent to net sown area is found in wide distribution in the Yashoda basin (Exhibit 6.5). One hundred and nine villages (70.32 per cent to total village) falls in this category. The area under hybrid jowar increases its areal extent towards east. The maximum area under this crop has been noticed at Sarur (43.61 per cent) and minimum at Kanchanpur (7.00 per cent), the maximum area it attributes

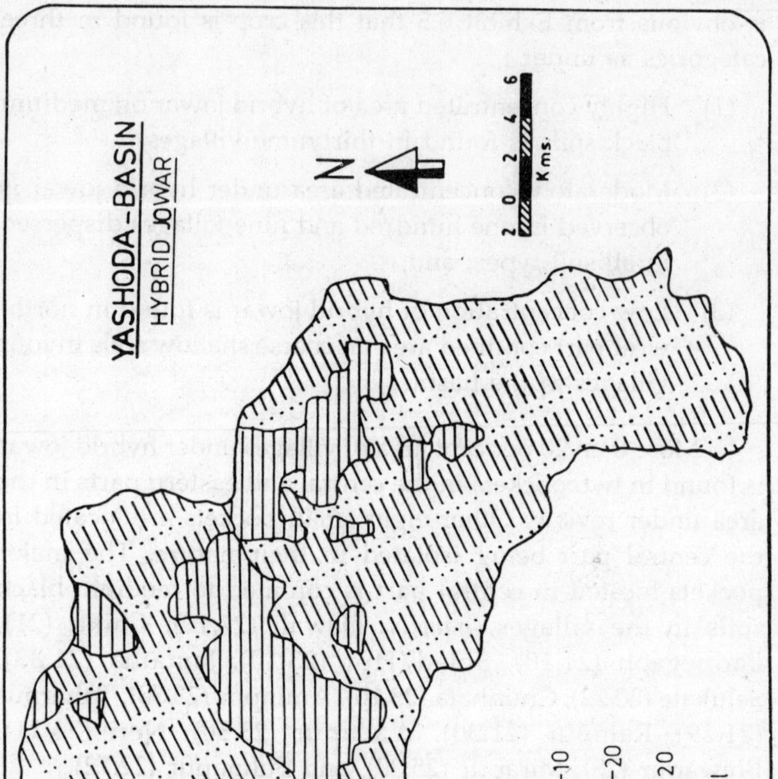
YASHODA-BASIN
HYBRID JOWAR
N
2 0 2 4 6
Kms.
LEGEND
0-10
10-20
> 20
Nil

EXHIBIT — 6.5

to the local fertile soils on which this crop is cultivated. The major concentration of hybrid jowar is found in the villages having more than 15 per cent area to net sown area, *namely*, Jawalgaon (17.3), Thangaddeo (15.18), Sirasgaon/Dnd. (13), Asala (19.13), Bhiwapur (15.29), Dapuri (17.64), Sonegaon/ Stn. (16), Itala (15.30), Dewangaon (16.35), Bhawanpur (16.66), Talegaon (15.32), Chikhali (17.11), Sonegaon/Bai (11.81), Adegaon (12.20), Raipur (16.57), Chinchala (15.38), Takali/ Chnj (19.38), Sekapur (18.32), Bhojankheda (14.85), Sarur (19.59), Mamdapur (17.19), Borgaon (18.32), Giroli (15.27), Allipur (16.4), Pawani (16), Sonegaon/Khr. (16), and Bhagawa (17.00 per cent) in the study area.

Less than 10 per cent to net sown area under hybrid jowar is confined to northwest part on shallow coarse soils (Exhibit 6.5) in the villages namely, Husenpur (3.30), Renukapur (2.19), and Bodad (9.15) in the Yashoda basin.

Local Jowar (Andropogen Sorghum)

Local jowar is staple crop provides food to human and fodder to livestocks. Due to its large in size, local jowar is known as Great Millet. This crop locally is called Ganathi Jowar. It is full stem and leafy growth. The dried fodder is locally known as "Kadaba" local jowar occupies on 8049.18 hectares (11.74 per cent to net sown area) and stands fourth in rank in the Yashoda basin.

Ecological Conditions and Farm Operation

Local jowar is essentially a tropical crop grown in both kharif and rabi season. It thrives better under the condition of high temperature of 27°C. It requires average temperature between 27°C and 32°C during growing period however, many varieties of local jowar bears minimum temperature between 16°C and maximum 40°C. It requires average annual rainfall between 600 and 1000 mms. Local jowar is drought resistant, remains dormant during the period of drought and starts active growth when the moisture in soils improve. Both excessive moisture and prolong drought are harmful to local jowar.

The traditional method is used for the preparation of field. The land for local jowar is prepared by ploughing in May by a mouldboard plough (both by wooden plough and iron plough) upto a depth of 15 to 20 cms. Moreover two to three cross-wise harrowing are given after commencing monsoon rainfall. Generally, Bardi Soils (coarse shallow soils) are preferred for sowing local jowar. Traditionally, local jowar seed is mixed with mung before sowing and is blended with Kerosine oil to avoid further diseases arises during germination period. Drilling method is common for sowing local jowar in the basin. This crop hardly needs chemical fertilisers and sprayings. Interculturing is done when local jowar is of three week old with a slit blade hoe and subsequent interculturing are given at the interval of about fifteen days, two to three interculturing are common till this crop is 40 to 50 days old thereby it helps in controlling weeds and loosening soils. The dried local jowar plants cut down by labourers. The threshed grains are winnowed, dried in sun and stored safely.

Spatial Distribution

The spatial distribution of local jowar is largely controlled by average annual rainfall, soil types and attitude of farmers of the study region. Increasing awareness among farmers to adopt new varieties of crops for sowing has changed the cropping pattern of local jowar and hence the major concentration of this crop is mostly confined to west and northwest parts in the Yashoda basin. The main features of spatial distribution of local jowar are as below :

(1) Highly concentrated area of local jowar where more than 30 per cent land to net sown area is found in seven villages

(2) Moderate concentration of local jowar having 20 to 30 per cent land to net sown area is observed in eighty seven villages

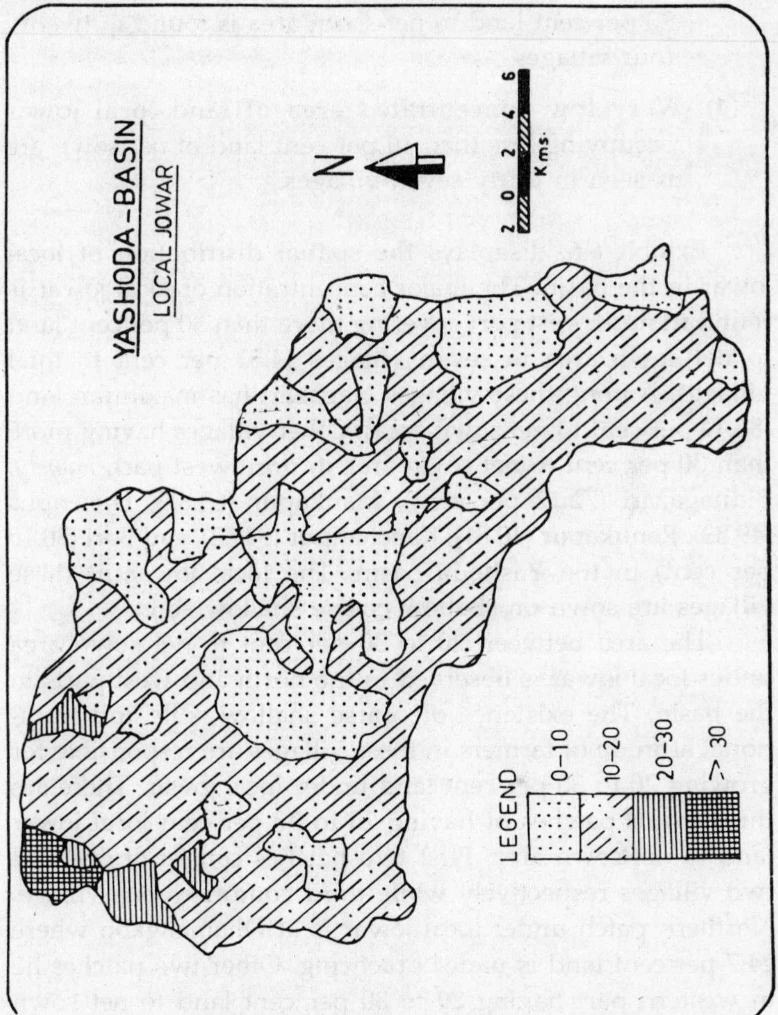
YASHODA-BASIN
LOCAL JOWAR
N
2 0 2 4 6
Kms.
LEGEND
0–10
10–20
20–30
> 30

EXHIBIT — 6.6

(3) Low concentrated area under local jowar where 10 to 20 per cent land to net sown area is found in twenty four villages

(4) Very low concentrated area of land local jowar occupying less than 10 per cent land ot net sown are is seen in thirty seven villages.

Exhibit 6.6. disaplays the spatial distribution of local jowar in the basin. The major concentration of local jowar is found in northwest part covering more than 30 per cent land to net sown area in seven villages (4.51 per cent to total villages. Among these villages Nagazari has maximum land (83.14 per cent to net sown area). Other villages having more than 30 per cent to net sown area in northwest part, *namely*, Bidnagazari (72.12 per cent), Mankapur (63.81), Husenpur (49.83), Renukapur (40.81), Kanchanpur (34.82), and Bori (30.10 per cent) in the Yashoda basin. The local jowar in these villages are sown on shallow coarse shallow soils.

The area between 20 to 30 per cent to net sown area under local jowar is observed in the north and west parts in the basin. The existence of coarse shallow soils and traditional attitude of farmers in these villages are responsible for growing 20 to 30 per cent land under local jowar. There are three major patches of having 20 to 30 per cent local jowar land to net sown area. First two patches consist of one and two villages respectively while third comprises one village. Northern patch under local jowar is at Dhamangaon where 24.7 per cent land is under practicing. Other two patches lie in western part having 20 to 30 per cent land to net sown area. The large patch in west consists of two villages, *namely*, Barasonegaon (28.15) and Laxminarayanpur (23.91), while third is confined to Sekapur where 20.52 per cent land is under local jowar. The cultivation of local jowar, in this area, is on coarse shallow soils where traditional ways of practicing agriculture is still prevalent.

About 10 to 20 per cent area to net sown area under local jowar exhibits wide distribution in the Yashoda basin. This category of local jowar is observed in central and southern parts on coarse shallow to medium deep black soil. In central part local jowar has dominant position producing high yield per hectare than that of western part in the basin. Maximum area of local jowar is observed at Meghapur (20.00 per cent to net sown area), while minimum at Talegaon (10.10 per cent to net sown area). The areal extent for cultivating local jowar is mainly dependent on the attitude of farmers of the village, besides types of soils which is ultimately resulting yields. The major dominant area of villages where 10 to 20 per cent land to net sown area is devoted to local jowar are *namely*, Lonsawali (1920), Dorali (19.00), Waifod (13.29), Kurzadi (12), Bodad (1'4.91), Amboda (13.62), Nimgaon (12.51), Kelapur (11.45), Kurla (15.22), Pimpri (15.50), Loni (15.10) and Nandpur (11.20), in the basin. Other notable villages having 10 to 20 per cent land to net sown area under local jowar in Taharpur (12.41), Siwapur (13), Ashta (12), Chitki (15.25), Malkapur (17.37), Karla (15.22), Pimpri (15.50), Khadaki (15.00), Muradgaon (13.34), Sonegaon/Abj (14.85), Dagapur (15.12), Hirapur (15), Akoli (15), Isapur (13.21), Krishnapur (15.95), Aloda (15.24), and Bhagawa (13) in the Yashoda basin.

The spatial distributon of local jowar in the basin occupying area less than 10 per cent to net sown area is found in patches in central, northeast and eastern parts (Exhibit 6.6). The minimum area of local jowar in this category is found at Banpur (5.61 per cent to net sown area) in central part, while maximum at Dighi (10 per cent to net sown area). The eight villages (5.16 per cent to total villages) have observed less than 10 per cent land under local jowar in central part in the area under review. These villages are *namely*, Paloti (9), Bhugaon (9.91), Asharatpur (8.85), Wallapur (9), Dapuri (9.98), Yesgaon (9.97), Sarur (8.13), and Nandpur (7.18) in the basin.

Wheat (Tritium Vulgare/satium)

Wheat is leading rabi crop in the Yashoda basin. It stands third in rank in the basin, occupying 8754.57 hectares (12.7) percent to net sown area. Wheat is direct source to man. It is cheaf substance of 50 per cent people in the world. It can be used for many purposes such as bread-making, chapati and in different forms as food.

Ecological Conditions and Farm Operation

Wheat grows in varying climatic conditions but it does thrive well under cool, dry and clear wheather. It requires cool temperature in early stage for proper tillering. Dry, sunny days and cool nights are helpful. Moreover, long duration of days and short nights are essential for optimum growth of plants. Temperature ranging between 7°C and 21°C finds better for wheat. It requires rainfall between 750 to 1100 mms for initial growth. Wheat grows on moisture retaining soils. The winter showers during growth period is helpful to wheat. Sufficient clear and dry climate suits wheat growth. Wheat can be cultivated on different soil types from coase shallow to heavy clayed soils. It grows well on well-drained loam and clay loam soils. Regur in the study region is suitable for getting high yield per hectare.

The preparation of the land on which wheat is to be grown, should be fine, compat and seedbed. One ploughing is essential with an iron plough for opening subsurface soil and keeping annual weeds under partial control. It is followed by four five harrowing during monsoon with Bakhar (a blade harrow) for weed control. One shallow bakharing and one light planking are essential just before sowing. The sowing of wheat starts usually in second half of October and continues till middle of November. Long dry days and cool nights in the basin support for the wheat growth. Wheat is rarely manured. It is, generally, harvested in February or March when grain is dead ripe and straw is golden yellow. The brittle plants are usually cut with a sickle and stocked in

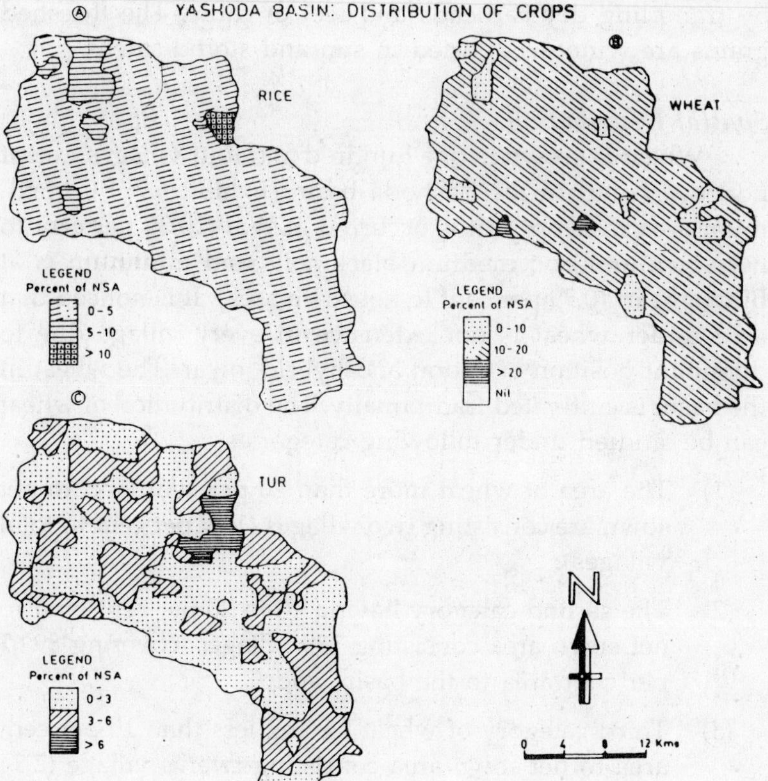

YASHODA BASIN: DISTRIBUTION OF CROPS
A
RICE
LEGEND
Percent of NSA
0-5
5-10
> 10
B
WHEAT
LEGEND
Percent of NSA
0-10
10-20
> 20
Nil
C
TUR
LEGEND
Percent of NSA
0-3
3-6
> 6
N
0 4 8 12 Kms

EXHIBIT — 6.7

the field to dry or only matured earheach. Threshing is done by trampling dry earheads and feet of cattle. The threshed grains are winnowed, dried in sun and stored safely.

Spatial Distribution

Wheat is grown in one hundred and four villages except northwest part in the Yashoda basin (Exhibit 6.7/B). Maximum area under wheat is occurred at Wai (20.39 per cent to net sown area) on medium black soils and minimum is at Bidnagazari (0.79 per cent to net sown area). It is noticed that area under wheat is not extensive in every village due to dominant position of cotton and hybrid jowar. The wheat in the basin is cultivated traditionally. The distribution of wheat can be studied under following categories :

(1) The area of wheat more than 20 per cent land to net sown are consisting two villages (1.29 per cent to total villages).

(2) The second category having 10 to 20 per cent land to net sown area consisting 139 villages, covering 89.67 per cent area to the basin, and,

(3) Third category of wheat having less than 10 per cent area to net sown area consisting twelve village (7.74 per cent to total villages) covering 8.38 per cent area to the basin.

The central and southern parts in the basin have more than 20 per cent area to net sown area (Exhibit 6.7/B) in two villages and moisture retaining *namely*, Wai (20.39) and Bopapur (20.17) on medium black soils. About 10 to 20 per cent land to net sown area is found wide distribution in the area under review. The area of wheat, generally, decreases towards west. The areal extent of wheat seems to vary from village to village. The western and northern parts of the basin have 10 to 15 per cent land to net sown area in twentyfour villages (15.48 per cent to total villages), *namely*, Bori (10.33),

Barasonegaon (12.92), Lonsawali (11.20), Dhamangaon (12.61), Wathoda (13.77), Śindi/Kh. (11.91), Amala (12.12), Tigaon (12.47), Umari (12.67), Waifad (13.72), Dahegaon (13.20), Rotha (13), Bodad (14.20), Degaon (13.25), Amboda (18.81), Sindi/Mgh (12.24), Kelapur (11.44), Nimgaon (13), Laxminarayanpur (15), Chitki (13.2), Sekapur (10.15), Mulkapur (12.30), Karala (10.57), Nandpur (10.45), and Loni (12.20). The areal extent between 10 and 15 per cent to net sown area under wheat is found confining to low rainfall zone (average annual rainfall of 900 mms).

In the south and eastern parts, have areas between 10 to 20 per cent land under wheat to net sown area are confined to fertile soils producing high yield per hectare in 37 villages (23.87 per cent to total villages). These villages are *namely*, Ashta (11), Jamnala (12.61), Bhiwapur (13.14), Dapuri (10.57), Dattapur (10.50), Sonegaon/Stn. (10.27), Ambapur (12.13), Talegaon (10.29), Sonpeth (12.50), Ajanavati (11.48), Ekurli (10.32), Adegaon (11.22), Raipur (13.21), Chichala (14.12), Takali/Drn. (10.11), Sekapur (17), Aloda (10.19), Sarur (10.02), Borgaon (18.20), Pathari (17.38), Nimsoda (12.41), Takali (12.32), Allipur (10.81), Rohankheda (13.82), Alamdoh (15.62), Gadegaon (12.60), Pawani (10.07), Chanki (12.66), Sonegaon/Knr. (14.39), Kachangaon (15.12), Bhagawa (10.31), Sirasgaon/Bzr (11.25), Mansawali (14.70), Nidha (13.54), Yarangaon (10.24), Kanholi (14.6), and Takali/Jjr. (13.41) in the study area under review.

The area under wheat having less than 10 per cent to net sown area is found in patches. Two dominant patches are observed in north west parts in four villages. The low area under wheat has been due to low fertility of upland hilly soils which produces low yield per hectare. These villages are *namely*, Bidnagazari (0.79), Mankapur (5.19), Husenpur (8.93), Kurzadi (10). Five patches of wheat having less than 10 per cent land to net sown are is observed in Padhegaon (7.11), Agargaon (9), Ajagaon (9.53), Jawalgaon (9.31), Neri (9), Dapuri (7.30), Bhiwapur (9.80), Dattapur (9.10) and Gioli (9) in the central part of the Yashoda basin. The central part has medium to deep black soils which are mostly used for cultivating hybrid jowar and cotton.

Tur (Cajancus Indicus)

Tur is kharif crop grown in the Yashoda basin. It accounts for 2.89 per cent to net sown area (1983.13 hectares) on one hundred and fiftyfive villages. It is considered that tur plant is basically of Africa, borrowed and distributed in almost all tropical countries. Tur is sown with cotton two rows of tur after every ten rows of cotton in the Yashoda basin.

Ecological Condition and Farm Operation

The humid climate is suitable for germination and early vegetative growth. Tur can be cultivated in all soil types, choice mostly depending on the attitude of the farmers but medium black soils resulted high yield per hectare. The average annual rainfall between 800 and 1100 mms favours proper growth. It requires presence of moisture in soils for healthy generation of tur plants. The occurrence of frost damages plants. Cloudy weather adversely affects and gives rise to pesticides. Tur needs temperature between 27 to 30°C for better growth. The higher temperature during maturity stage supports for early drying.

The separate land preparation for tur crops is not practiced in the basin as it is sown in coton field. Tur, like cotton field, requires two-three ploughing for smoothing soils. It is sown generally in second week of June with cotton. The usual practice is to sow it in two rows after every ten rows of cotton. It needs six to nine kilograms tur for one hectare. Three to four harrowings are given to tur for its proper growth. Three sprayings are essential to control pesticides. After maturity of tur it is cut down by means of sickles and dried in sun. Threshing is done by beating sickle. Grains are separated from chaft by winnowing. Leaves and broken bodes serve as cattle feed and stalks are used for basket making and thatching material.

Spatial Distribution

Tur ranks sixth accounting for 2.89 per cent to net sown area for 1985-86 in the Yashoda basin. It is mainly produced for local consumption and therefore hectarage under tur has less in every villages in the basin. The maximum area of tur is found at Sawangi/Mgh (8.33 per cent to net sown area) in northeastern part whereas minimum area under tur is at Sarur (1.16 per cent to net sown area). The percentage of tur in the basin generally decreases towards west from east (Exhibit 6.7/C). This hectarage is mainly depended on the cotton hectarage as tur is mixed with cotton. The spatial distribution of tur in the Yashoda basin can be studied under three classes as under :

(1) Areas having more than 6.00 per cent land under tur. It occurs in four villages in the basin.

(2) Land between 3 and 6 per cent and is found in fifty villages, and,

(3) Less than 3 per cent area under tur has wide distribution and is found in ninetyseven villages.

The major concentation of tur is found in northeast part. This part has more than 6.00 per cent land to net sown area under this crop (Exhibit 6.7/C). This area receives average annual rainfall between 700 and 900 mms. Four villages have observed more than 6 per cent to net sown area in this category. These villages are Sawangi/Mgh (8.33), Malegaon (7.13), Paloti (6.20), and Jamatha (6). About 3 to 6 per cent land under tur is found in fiftyfour (34.83 per cent to total villages) in the Yashoda basin which are observed in many patches. Among them five patches are visible covering 70 per cent in the area under review. First patch of four villages is located in northern part on coarse shallow soils which include Waifad (3.32), Sindi/Kh (3.10), Dahegaon/Mskn (3.69) and Amala (3.65). Second patch covers Dewali (3.11). Third patch of tur consists of two villages *namely,* Jamnala (4.17) and Selukate (4.61) in east part in the study area. Fourth patch having 3 to 6 per cent land to net sown area is found in eastern part on deep black soils where average annual rainfall ranging from 1000 to 1100 mms in seven villages (4.51 per cent to total villages). These villages are *namely,* Allipur (3.18), Babapur (3.91), Pawani (3.12), Kachangaon (3.19), Sirasgaon (3.9), Chanaki (3.55) and Mansawali (3.14) and Sixth major patch is located in central part on medium deep soils in seven villages (4.51 per cent to total villages). These villages are *namely,* Ashta (3.19), Siwapur (3.20), Dattapur (3.17), Bhiwapur (3.84), Dapuri (3.31), Ambapur (3.16), Taharpur (3.15) and Rampur (3.10) in the basin.

The third category of less than 3 per cent land to net sown area under tur found in ninetyseven villages (62.58 percent to total villages) in the Yashoda basin and their location is found throughout the region, despite soil types and average annual rainfall. The notable villages where tur is cultivated between 2 to 3 per cent out to net sown area are *namely,* Husenpur (2.33), Lonsawali (2.18), Dhumangaon (2.37), Wathoda (2.92), Tigaon (2.10), Kurzadi (2.11), Rotha (2.90), Degaon/tn. (2.17), Neri (21.10), Amboda (2.94), Dhotra (2.11), Sindi/Mgh (2.57), Chitki (2.10), Padhegaon (2.77), Solad/Hpr (2.90), Afzalpur (2.90), Banpur (2.13), Wathoda (2.90), Aminpur (2.90), Jawalgaon (2.90), thangaddeo (2.30), sirasgaonn/Dnd (2.10), Ajagaon (2.21), Asata (2.18), Mirapur (2.35), Sonegaon/Stn. (2.90), Itala (2.80), and Dewangaon (2.13) in the area under review.

Rice (Oryza Sativa Linn.)

Rice ranks fifth in the Yashoda basin. It accounts for 3.43 per cent to net sown area (2351.18 hectare). The rice is mainly cultivated for local consumption. Many varieties of rice are available and hence it is preferred by farmers according to local conditions.

Ecological Conditions and Farm Operation

Rice is mainly a tropical crop. It thrives well under high temperature and humidity. The distribution of rice depends on climatic conditions than soil types. The temperature ranging between 20°C and 37°C is optimum for growth. It grows well in the area where rainfall is between 1000 mms and 1100 mms. However, many varieties of rice can grow under rainfall as low as 400 mms and as high as 5000 mms. Laterite soils can be suitable for rice cultivation. Even then different varieties of soil may grow in coarse shallow and loam soils.

Field preparation generally commences in April and May with ploughing land by iron plough. Repeated harrowings are essential for obtaining good tilth. The selected varieties of rice are sown by drilling method in the basin in second week of June. Two to three interculturings are done within an interval of fortnight. It mainly, depends on the variety chosen and field condition. Interculturing is stopped before one month of rice harvesting. N.P.K. are given three to four times to maintain fertility and high yield per hectare. Rice is harvested with a sickle by manual labourers. It is threshed by sticks or under the feet of bullocks.

Spatial Dstribution

The spatial distribution of rice in the basin is shown in Exhibit 6.7/B. Rice cultivation is mostly depended on the farmers attitude. Exhibit 6.7/B reveals three distinct areas of rice distribution are as under :

(1) Areas having more than 10 per cent to net sown area.

(2) About 5 to 10 per cent rice percent to net sown area in seven villages and,

(3) The third area of having less than 5 per cent is under rice to net sown area is found in one hundred and fortyseven villages.

The land more than 10 per cent to net sown area under

rice is found in north eastern part in the Yashoda basin (Exhibit 6.7/B) at Sawangi/Mgh (10.30 per cent to net sown area). About 5 to 10 per cent land to net sown area is found in four patches in northwestern and western parts in the region. The northwestern villages have coarse shallow soils, producing low yield per hectare. The rice is here cultivated in kharif season. These villages are *namely*, Lonsawali (5.41), Waifad (5.18), Kanchanpur (6) and Barasonegaon (5.7 per cent) in the basin, while in western part at Kelapur (5.12) show low percent of land under rice. Other isolated patches of rice are found in central part of Palasgaon (5.12) and Shampur (7.17) in the basin.

Less than 5 per cent land under rice was observed in one hundred and fortyseven villages (94.83 per cent to total villages) covering large part in the basin (Exhibit 6.7/B). The hectarage under this crop decreases towards west as the soil texture decreases. The villages having low hectarage under rice due to dominance of hybrid jowar, cotton and wheat. The central part has following villages where less than 5 percent land is under rice, *namely*, Nagthana (4.31), Bhaiyapur (3.91), Chikani (4.11), Hirapur (3.13), Selukate (3.35), Babapur (3.91), Kurzadi (3.64), Aminpur (3.59), Wallapur (4.11), Jawalgaon (3.22), Thangaddeo (4.29), Narayanpur (3.29), Sirsgaon/Dnd (3.78), Asata (3.51), Siwapur (4.12), Ashta (4.34), Jawnala (4), Dattapur (3.50), Dewangan (4.11), Dewali (5.00), Chikkali (3.50), Sonegaon/Bai (3.11), Rapur (3.91), and Aloda (3.15) in the Yashoda basin.

Linseed (Linum Uṣitatissimum Linn)

Linseed is one of the oilseeds grown in the basin on 1262.15 hectares (1.84 per cent to net sown area. It is locally known as "Jowas" or Alshi. This crop ranks twelve in the area under review.

Ecological Conditions and Farms Operation

This crop can grow in all soil types and hence the yield varies. Well-drained medium black soils generally favour proper growth of linseed. It thrives well in cool, dry and clear weather conditions. Long duration of days and short nights are essential for the optimum growth. Linseed requires temperature between 7°C and 21°C but it can be cultivated even in temperature of 28°C. Average annual rainfall between 750 and 1000 mms is essential. The occurance of rainfall in December and January prove to be helpful for proper growth.

Linseed field requires careful landing. One ploughing is done after every three years with an iron inversion plough for opening sub-surface. It is generally sown in October and is harvested after 100 days, when linseed begins to dry up turning brownish by pulling or by cutting close to the ground and stocked on thresting floor. It is threshed under the feet of bullocks or with the help of sticks.

Spatial Distribution

Linseed is a rabi crop sown on 1282.15 hectares. It ranks twelve in the basin. It is observed that the low hectare under this crop is due to dominanace of cotton, hybrid jowar and wheat which have comparatively high yield per hectare in the Yashoda basin. (Exhibit 6.7/A). Maximum land under linseed is observed at Husenpur (5.16 per cent to net sown area) while minimum is Solad/Hpr (0.19 per cent to net sown area). The distribution of linseed in the basin shows as under :

(*1*) Area of more than 3 per cent to net sown area. It is found in twelve villages, and

(2) Low hectarage where less than 3 per cent land to net sown area occurring in one hundred and forty villages.

It is observed that more than 3 percent land to net sown area has concentrated in western, central and southern parts in the basin in many patches. The largest patch of linseed is found in west part consisting of four/villages (2.58 per cent to total villages) on coarse shallow soils. These four villages are, Malkapur (3.30), Nandpur (3.17), Muradgaon (3.43) and Khadaki (3.11). Other patches of this category are isolated and distributed throughout the basin covering villages, *namely*, Rohankhea (3.31), Gadegaon (3.16). In the central, these villages have more than 3 per cent land under linseed. Those villages are *namely*, Palasgaon (3.19), Sirasgaon/Dnd (3.14), and Takali (Jjr) (3.12).

Less than 3 per cent land to net sown area is found in one hundred and forty villages (90.32 per cent to total villages). The distribution of this category varies with soil types and rainfall amount. The maximum hectarage under this category is observed in eastern part where cotton and hybrid jowar are dominant in hectarage at Mirapur (3.92 per cent), whereas minimum hectarage is found at Salod/Hpr (0.10) in the central part in the basin. The significant land in the villages under this crop are found at Lonsawali (2.82), Dorali (2.11), Wathoda (2.17), Dahegaon/Msakn (2.11), Sindi/ Mgh (2.35), Asala (2.62), Kurzadi (2.87), Rampur (2.73), Neri (2.10), Dattapur (2.13), Sonegaon/Bai (2.91), Chinchala (2.17), Bhajankheda (2.18), Sarur (2.23), Mansawali (2.10), Takali/ Drn. (2.65) and Pawami (2.28), in western, central and southern parts on coarse shallow to deep black soils in the Yashoda basin.

Mot (Phaseolus Aconitifolius Jaca)

Mot is kharif crop occupying 1551.06 hectares (2.26 per cent to net sown area. It holds eighth position among crops grown in the Yashoda basin.

Ecological Conditions and Farm Operation

Mot is grown on medium deep black soils in the basin but it can be grown on all soil types. Its yield per hectare depends on soil types used for mot cultivation. The average

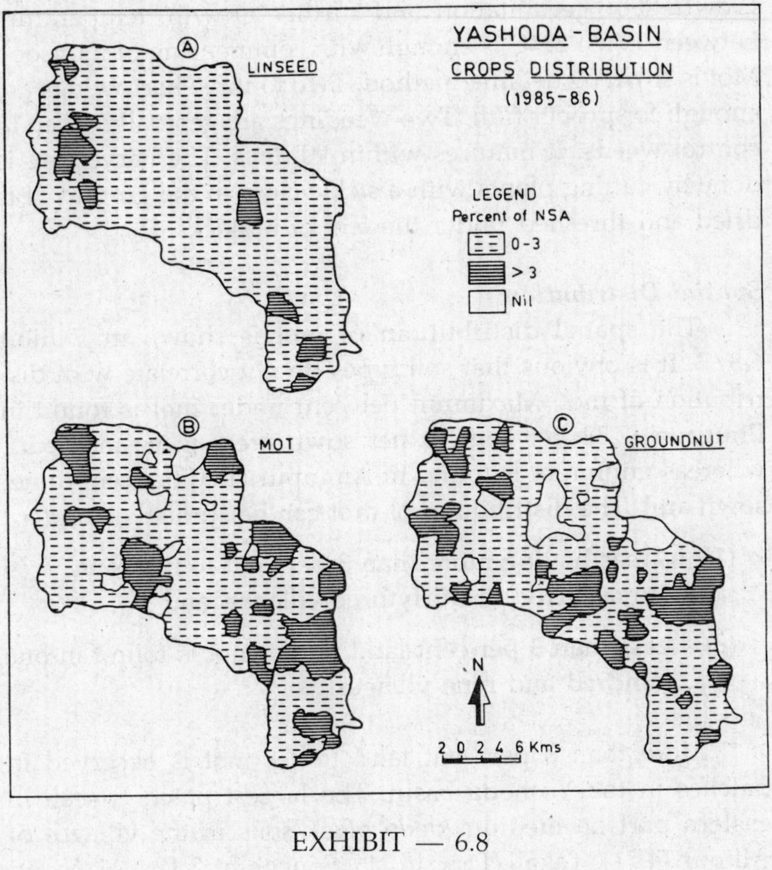
YASHODA - BASIN
CROPS DISTRIBUTION
(1985-86)
A
LINSEED
B
MOT
C
GROUNDNUT
LEGEND
Percent of NSA
0-3
> 3
Nil
N
2 0 2 4 6 Kms

EXHIBIT — 6.8

annual rainfall between 650 and 100 mms is suitable for proper growth. For germination and further growth, temperature between 15° to 21°C is enough with commencing of monsoon Mot is sown by drilling method. Two to three harrowings are enough for proper tilth. Two weedings are generally given to control weeds. It matures within 90 days. The harvesting is done by cutting plants with a sickle close to the ground. It is dried and threshed under the feet of bullocks.

Spatial Distribution

The spatial distributioin of mot is shown in Exhibit 6.8/B. It is obvious that soil types do not correlate with distribution of mot. Maximum per cent under mot is found in Bhugaon (7.32 per cent to net sown area) in .eastern part, whereas minimum is found in Aminpur (1.00 per cent to net sown are). The distribution of mot can be studied as under :

(1) Areas having more than 3 per cent to net sown area are occurred in thirtythree villages, and

(2) Less than 3 per cent land under mot is found in one hundred and nine villages.

More than 3 per cent land under mot is observed in patches in the Yashoda basin. The largest patch is seen in eastern part on medium deep black soils in the villages of Allipur (4.71), Takali/Drn. (4.21), Sonepeth (3.12), Ajanavati (3.10) and Ekurli (3.12) in the area under review. Second patch is observed in northwest part producing mot on coarse shallow soils in seven villages (4.51 per cent to total villages), *namely*, Bid-nagazari (7.09), Mankapur (5.11), Husenpur (5.17), Renukapur (6.10), Kanchanpur (4.32), Bari (3.52), and Barasonegaon (3.11). Third patch of mot is found in central part in the villages of Selukate (3.55), Bhugaon (7.32), Chunhala (3.19), Aminpur (3.20), and Taharpur (3.90). The western patch under mot has been found in two villages *namely*, Yesgaon (3.19), and Neri (3.12).

The wide distributioin of mot is found in one hundred and nine villages (70.32 per cent to total villages) through out the basin. The distribution of mot is largely influenced by rainfall amount and soil types. It has been observed that low fertility land is generally used for mot cultivation. In this category maximum mot area in per cent is found at Gadegaon (2.98 per cent to net sown area) in the eastern part whereas minimum is at Rotha (0.17 per cent to net sown area) in northern part in the basin. The major concentration of mot is observed in this category in western and northern part in the villages of Nagazari (2.7), Lonsawali (2.7), Amala (2.11), Waifad (2.08), Dhotra (2.47), Sind/Kh. (2.35), Kelapur (2.11), Padhegaon (2.01), Paloti (2.94) in the Yashoda basin.

Groundnut (Arachis Hypogaea Linn)

Groundnut is a cash crop grown in kharif season in the Yashoda basin on 1483.09 hectares (2.14 per cent to net sown area). It is rainfed crop. It maintains fertility and reduces the cost of preparaion for succeeding crops. The groundnut is spread out from South America to all over the world.

Ecological Conditions and Farm Operation

Groundnut is rainfed crop grown in tropical and sub-tropical location. Generally, groundnut can grow well and thrive in long and warm growing season temperature rainging between 21°C and 26°C. It does not withstand waterlogging and drought conditions. This crop requires rainfall between 625 to 1000 mm. Heavy and prolonged rainfall shunts the growth. Groundnut can grow successfully on sandy loam, loam and well-drained soils. It requires a uniform, fine and loose seedbed. The deep ploughing in May is followed by harrowing in summer is essential. This crop is sown in June. It needs hand weeding and one or two bullock hoeing three weeks after sowing. Groundnut is harvested in September-October by uprooting plants when the lower leaves turn yellow and start withering and pods are well developed and tough to feel. The uprooted plants are separated from pods, dried in sun and get ready for sale.

Spatial Distribution

Exhibit 6.8/C shows spatial distribution of groundnut in the Yashoda basin. One hundred and fortythree villages (92.25 per cent to total villages) are cultivating this crop accounting for 2.14 per cent to net sown area. The hectarage under groundnut decreases from west to east. Maximum hectarage is observed at Renukapur (5.51 per cent to net sown area) in northwestern part on coarase shallow soils, while minimum is at Dhamangaon (0.23). The distribution of this crop has found in many patches and disintegrated throughout the basin.

More than 3.00 per cent land to net sown area is observed in forty villages (25.80 per cent to total villages) in three patches in western and eastern parts on coase shallow to medium black soils in the Yashoda basin. These three patches are found where mug and mot do not have extensive hectarage. The major patch of concentration of more than 3.00 per cent is confined to western part in three villages *namely*, Kurla (3.12), Nandpur (3.10), and Khadaki (4.45). The eastern patch of this category is seen in eight villages that include Sonegaon (3.77), Sultanpur (3.55), Talegaon (3.77), Ajanavati (3.26), Sonpeth (3.95), Ekurdi (3.21), and Bhavanpur (3.39) whereas central patch where more than 3.00 per cent land under groundnut is found in four villages, *namely*, Dighi (3.26), Bopapur (3.50), Sirasgaon/Dnd (3.15), and Sonegaon/Bai (3.17) in the Yashoda basin. Other thirteen patches are small in size have distribution under groundnut where more than 3.00 per cent to net sown area is found under this crop. These patches are located in all parts in the basin. The notable patches are at Bara-Sonegaon (3.50), Dorali (3.19), Kurzadi (4.00) Degaon/Stn. (3.12), Dattapur (3.33), Borgaon (3.11), and Sonegaon/Knr (3.64) in the area under review.

It is observed from Exhibit 6.8/C that the land occupied

less than 3.00 per cent was found in extensive areal extent in one hundred and three villages (66.45 per cent to total villages) on all soil types throughout the basin. The distribution of this crop does relate to rainfall amount and soil types. The villages where hectarage between 2 to 3 have been observed are *namely*, Bidnagazari (2.11), Husenpur (2.11), Amala (2.42), Tigaon (2.42), Bodad (2.67), Nagthana (2.11), Meghapur (2.81), Banpur (2.99), Kesalapur (2.90), Kurzadi (2.47), Jawalgaon (2.50), Taharpur (3.00), Ashta (3.00), Jamnala (2.10), Bhiwapur (2.31), Sultanpur (2.50), Mirapur (2.71), Dapuri (2.50), Ambapur (3.00), Dewangaon (2.31), Yesgaon (2.39), Jamani (2.54), and Pimpri (2.81) in the Yashoda basin.

Mung (phaseolous Radiatus Roxb)

Mung is kharif crop occupying 1611.46 hectares (2.25 per cent to net sown area) in the basin. Mung grains are used as pulses. The husk of mung is used for cattle feeding.

Ecological Conditions and Farm Operation

Mung thrives well on medium to deep black soils. This crop requires 21° to 30°C temperature and rainfall between 600 to 750 mms. Heavy rainfall during flowering cause harm to mung. Generally, mung field is prepared by one ploughing and twice harrowng to obtain rough tilth. It is sown in second week of June and harvested in October. Mung is mixed with hybrid jowar and local jowar in the basin. Drilling method is common for sowing. Two weedings are essential, one after 20 days sowing. It matures within 70 days. Harvesting is done by labourers with sickles. It is dried and later threshed by trampling with sticks.

Spatial Distribution

The mung is grown in one hundred and forty nine villages (96.12 per cent to total villages) throughout the basin. The major concentration of mung is observed in the area where hybrid jowar and local jowar are dominated in the region as it is mixed sown with these crops. Maximum hectarage has been recorded at Bidnagazari (13.38) on local jowar dominating area in west part, while minimum is registered at Hirapur having 0.12 per cent land under this crop in central part (Exhibit 1.09). The mung distribution in the basin can be studied as under :

(*1*) Areas having more than 6.00 per cent to net sown area under mug. It is found in six villages.

(2) About 3 to 6 per cent land under mung has found in thirtythree villages, and

(3) Less than 3.00 per cent hectarage is on hundred and ten villages.

It is observed that more than 6.00 per cent land under mung occurs in the northwest part where local jowar is grown on coarse shallow soils and mung has low yield per hectare. The villages include *namely*, Nagazari (12.13), Bid-nagazari (13.38), Mankapur (13.11), Husenpur (7.60), Renukapur (8.22), and Bori (9.32).

About 3 to 6 per cent land under mung has observed in many patches is southern, central and northern parts in the basin. Maximum per cent of this category is observed at Takali/Chni (5.56 per cent to net sown area) in central and minimum at Chanki (3.10 per cent to net sown area) in the southern part in the basin. The largest patch of mung cultivation is found in the southern part where it is sown with hybrid jowar in the villages *namely*, Gadegaon (3.16), Nidha (3.16), Kanholi (3.15), Takoli (3.31), Yarangaon (4.31), Kachangaon (3.21) on medium deep black soils where mung is mixed sown with hybrid jowar.

Less than 3.00 per cent land under mung is found in all parts in the basin (Exhibit 6.9). This crop is grown, here, despite soil types. This category having maximum hectarage is at Asala (3.00 per cent to net sown area) and minimum recorded at Hirapur (0.12 per cent to net sown area). Most of the

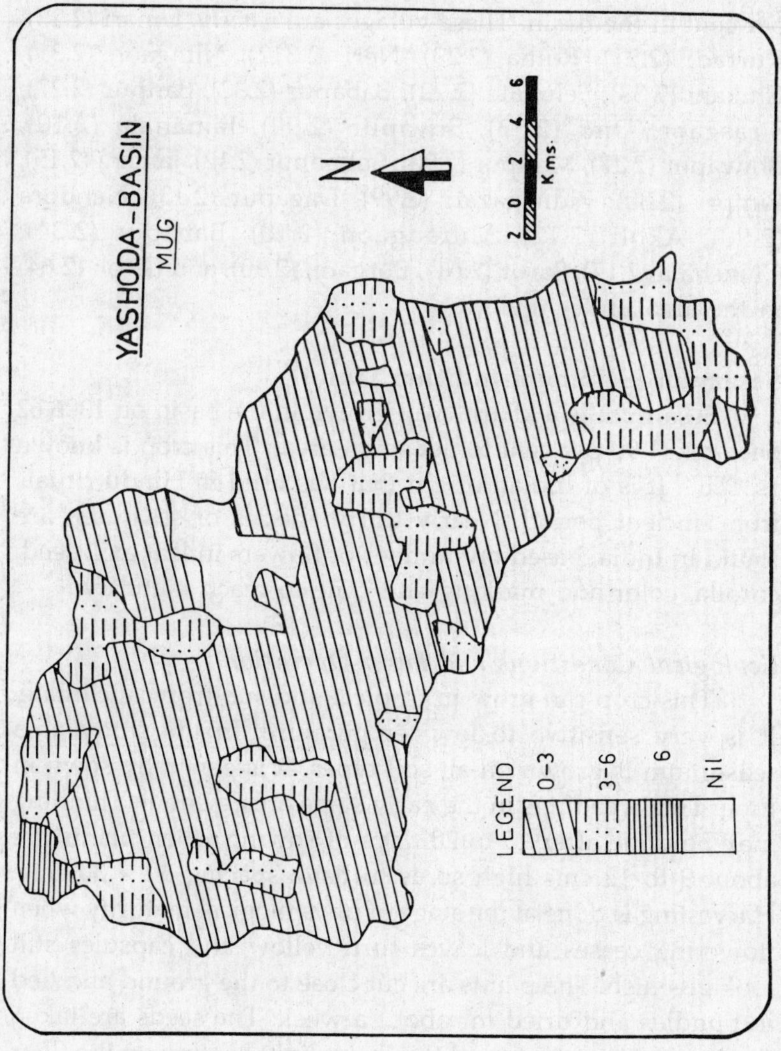

EXHIBIT — 6.9

villages in this category has observed to cultivate land 2 to 3 per cent in the Basin. These villages are *namely*, Umari (2.27), Kurradi (2.21), Rotha (2.23), Neri (2.312), Nimgaon (2.31), Chikani (2.33), Selukate (2.22), Babapur (2.32), Banpur (2.27), Sirasgaon/Dnd (2.52), Siwapur (2.90), Jamanala (2.10), Bhiwapur (2.27), Mirapur (2.90), Sultanpur (2.12), Jamani (2.15), Pimpri (2.80), Bidnagazari (2.99), Dagapur (2.13), Nandora (2.90), Akoli (2.12), Muradgaon (2.10), Babapur (2.34), Chinchala (2.17), Sarur (2.61), Borgaon (2.90), and Giroli (2.64) in the area under review.

Sesamum (Sesamum Indicum Linn.)

Sesamum is a kharif crop grown in the basin on 1306.62 hectares (1.90 per cent to net sown area). This crop is known as "Til". It is of Indian origin that has used in Hindu rituals from ancient period. Nearly thirty species of sesamum are found in India, based on number of flowers in leat axil, seed, corolla, color and marking and time to reach maturity.

Ecological Conditions and Farm Operation

This crop can grow in tropical and sub-tropical climate. It is very sensitive to low temperature. In the study area seasumum is grown on all soil types. It is generally sown in June. Two weedings in the early stage are essential. Thinning may be done after 15 to 20 days of sowing when plants are about 9 to 12 cms high so as to have spacing 30 × 30 cms. Harvesting is done at the stage of physiological maturity when flowering ceases and leaves turn yellow and capsules still look greenish. The plants are cut close to the ground and tied in bundles and dried for about a week. The seeds are taken out by shaking plants and finally by light beating on the floor or on a piece of cloth for taking out total seeds.

Spatial Distribution

It is observed from Exhibit 6.10/A that the Yashoda basin has cultivated sesamum on one hundred and forty three villages (92.25 per cent to total villages). The maximum hectarage under sesamum is found in northwest part on coarse shallow soils at Renukapur (7.73 per cent to net sown area), whereas minimum hectarage of this crop recorded at Dhamangaon (0.11 per cent to net sown area). The remarkable feature of spatial distribution of sesamum is that the major concentration has been noticed in northern part, confining to coarse shallow soil. This fact explains that fertile soil is generally not brought under sesamum. The villages located in extreme northwest part have not found cultivating this crop in twelve villages (7.74 per cent to total villages) due to farmers' attitude to choose food crop, namely, loal jowar for their livelihood.

The spatial distribution of sesamum can be studied as under :

(*1*) Area having more than 3.00 per cent to net sown are has found in patches in thirteen villages and

(2) The sesamum hectarage is found less than 3.00 per cent. It is found in one hundred and thirty villages.

More than 3.00 per cent to net sown area under sesamum is shown in Exhibit 6.10/A in the Basin. The patches of sesamum are found in western, central and southern parts in the basin confined mostly to coarse shallow to medium soils. The notable villages where more than 3.00 per cent land under sesamum are found in the villages, namely, Padhegaon (3.52 per cent), Paloti (3.11), Kurzadi (4), Kesalapur (3.12), Dapuri (3.15), Basi and Takali (3.51) in the basin. Other category of land occupied by sesamum having less than 3.00 per cent to net sown area is noticed in one hundred and forty villages. The hectarage between 2 to 3 per cent land is found in the villages *namely*, Bara-Sonegaon (2.17), Dorali (2.31), Sindi/Kh (2.15), Waifad (2.90), Dahegaon (2.33), Sindi/Mgh (2.11), Padhegaon (2.41), Banpur (2.11), Waddha (2.21), Kurzadi (2.53), Wallapur (2.12), Jawalgaon (2.25), Thangadheo (2.10),

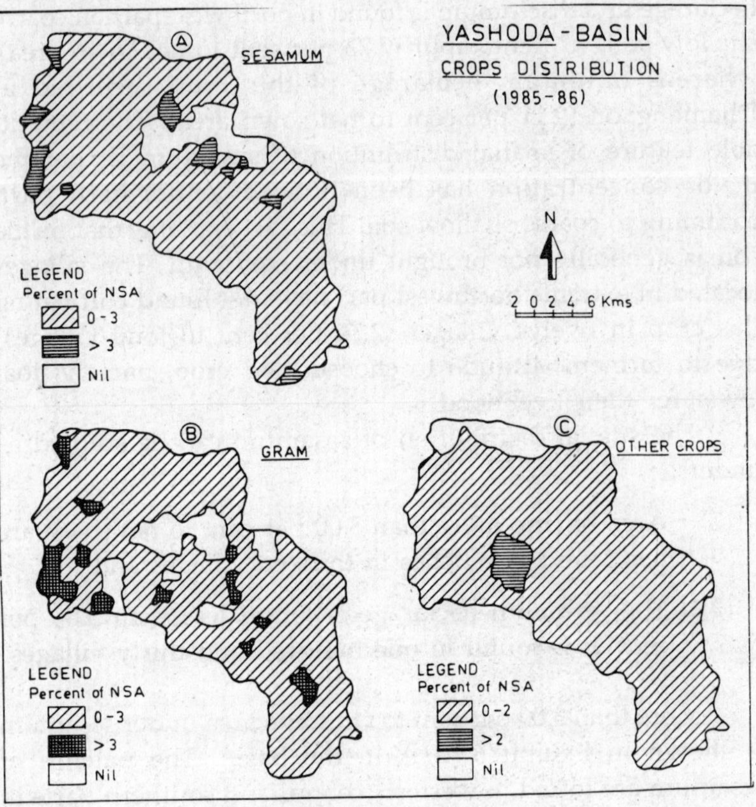
YASHODA - BASIN
CROPS DISTRIBUTION
(1985-86)
A
SESAMUM
N
2 0 2 4 6 Kms
LEGEND
Percent of NSA
0-3
> 3
Nil
B
GRAM
LEGEND
Percent of NSA
0-3
> 3
Nil
C
OTHER CROPS
LEGEND
Percent of NSA
0-2
>2
Nil

EXHIBIT — 6.10

Waigaon (2.11), Asala (2.12), Jamnala (2.11), Mirapur (2.10), Dattapur (2.72), Sultanpur (2.25), Sonegaon/Stn. (2.10), Mallkapur (2.90), Nandpur (2.90), Khadki (2.51), Belgaon (2.81), Dagapur (2.91), Nandra (2.71), and Isapur (2.90) in the study area under review.

Gram (Cicer Arietinum Linn.)

Gram is a rabi crop in the basin. It occupies 1270.92 hectares area under this crops (1.85 per cent to net sown area and holds eleventh rank in the region. It is used in various forms *i.e.* Dal (splits), floor, crushed or whole gram, cooked gram, salted gram, boiled or parched, roasted or cooked gram. Crushed grains and husk is used for livestock feeding.

Ecological Conditions and Farm Operation

This crop requires cool temperature and low rainfall between 600 and 700 mms. Heavy rainfall, occurance of hailstorms at the riping time cause damage the crop. Temperature ranging between 10° and 25°C is suitable. The moisture retaining clay loam and black soils are considerably fruitful for well growing. One ploughing and two harrowing are essential for sowing. Gram is sown in mid October by drilling method. This crop is rarely weeded or intercultured. It requires 100 days to turn yellow and dry. Harvesting is done in January with sickles cut close to the ground. After it is dried for a week in sun and later threshed under the feet of bullocks.

Spatial Distribution

Exhibit 6.10/B reveals spatial distribution of gram in the Yashoda basin. The spatial distribution of this crop is attributed to soil types and its moisture retaining capacity. It is observed that one hundred and fortyseven villages (94.83 per cent to total villages) are cultivating this crop. The maximum hectarage under growth is found at Alamdoh (9.21 per cent to net sown area) while minimum recorded in northwestern part at Yesgaon (0.22 per cent to net sown area) in the Yashoda basin. More than 3.0 per cent land under gram is occurred in twenty villages (12.90 per cent to total villages) in western, central and southern parts, confining to medium black soils. These villages are *namely*, Bidnagazari (4.61), Mankapur (3.22), Husenpur (8.11), Renukapur (8), Kanchanpur (4.90), Barasonegaon (3.73), in northwestern parts, whereas the villages having western location are of Malkapur (3.79), Loni (3.10), Nagazari (3.79), Timapur (4.15), Akoli (3.11), Muradgaon (4.70) and Ajanavati (3.50), Banpur (4.75), Ajagaon (6.1), Jamnala (6.90), Sekapur (4.27), Mamdapur (4.31), and Alamodh (9.21) in central part in the study area.

The villages having less than 3.0 per cent land under gram has wide distribution throughout the basin in one hundred and twenty seven villages (81.93 per cent to total villages). It is destributed dispite soil types and rainfall as gram require cool temperature and moisture retaining soils for its growth. Maximum hectarage under this crop is at Allipur (2.90 per cent to net sown area) while minimum recorded at Yesgaon (0.22 per cent to net sown area) in the basin. The major concentration of gram under this category has observed in south part in the villages *namely*, Sirasgaon (1.32), Mansawali (1.22), Nidha (1.33), Kanholi (1.11), Chanki (1.71), Yaransgaon (1.30), Kachangaon (1.15), Pawani (1.32), Gadegaon (1.51), Rohankheda (1.22), and Mendukdoh (1.35) in the study area on medium deep black soils.

Other crops

Other crops include Kultha, Tomato and Chillis. The basin show insignificant hectarage under this category (1.47 per cent to net sown area) occupying 950 hectarages and holds thirteenth rank among crops in the Yashoda basin.

Ecological Condition and Farm Operation

The above mentioned crops can grow in varied climatic conditions and soil types. Generally, these crops are cultivated in kharif season 700 to 900 mms rainfall is enough for growing these crops. The field preparation is similar to crops grown in kharif season in the basin. Two ploughings and three harrowings are essential. The spraying is needed to chillies and tomato.

Spatial Distribution

It is observed that one hundred and forty villages have less than 2.00 per cent land under other crops to net sown area (Exhibit 6.10/C). Dewali has recorded maximum area (3.50 per cent to net sown area) in the basin. Other crops particularly vegetables, chilli and tomato are cultivated on medium to deep black soils. This crop is raised on insignificant hecterage due to dominance of hybrid jowar and cotton in this part of the study area. The production of these crops are locally consumed. The land less than 2.00 per cent under other crops is distributed in all parts of the region but major concentration is found in central, eastern and southern parts in the villages, *namely*, Allipur (1.23), Takali/Drn (1.11), Rohankheda (1.10), Pawani (1.30), Gadegaon (1.23), Alamdoh (1.71), Chanki (1.40), Nidha (1.43), Mansawali (1.19), Yarangaon (1.20), Kanholi (1.31), Takali/Jjr (1.21), Pathari (1.51), Giroli (1.32), Borgaon (1.35), Aloda (1.11), Talegaon (1.32) and Sirasgaon/Bzr (1.50) in the basin while less than 2.00 per cent land under this crop is found in northwestern part where mainly Kultha is grown on coarse shallow soils in the villages *namely*, Nagazari (1.30), Bidnagazari, Laxminarayanpur (1.15), Bori (1.32), and Barasonegaon (1.17), in the area under review.

RESUME

The distributional pattern of crops in the basin has been described and interpreted taking into account the prevalent physio-socio-economic conditions in the basin and established cropping pattern. Cotton, local jowar, hybrid jowar, rice, mung, mot, linseed, groundnut, gram are grown (Table 6.3) in the basin. Among these crops, cotton and hybrid jowar are successfully grown and have relatively greater percentage under cultivation, amounting to 38.50 per cent and 16.88 per cent to net sown area respectively. Generally, both crops are grown and produced on medium and deep black soils in the basin owing to their productive nature. Cotton is mostly concentrated on deep black soil tract in the lower Yashoda basin and, therefore, this tract is reckoned to be highly productive part in the basin. The maximum area under cotton is observed at the village Sarur (43.61 per cent to net sown area). Hybrid jowar has been replaced by local jowar after 1975 due to his high yield per hectare having maximum percentage of cultivated land at the village of Palasgaon (39.59 per cent to net sown area). Other crops are negligible. The local jowar has 11.74 per cent land to net sown area, wheat (12.70 per cent), mug (2.35 per cent), mot (2.26 per cent) in the basin. The kharif period is important season from agricultural point of view when nearly 70 per cent area is under crops. Wheat is grown as a major crop in rabi season with its outstanding significance. Furthermore, the recent remarkable change is noticeable in the attitude of farmers to prefer high yielding crop varieties than those of traditional low yielding ones. This is a progressive welcome change in improving agricultural landuse pattern in the basin. One comes across, this change is observe mainly in the central and the eastern parts on fertile tracts in the Yashoda basin.

To comprehend meaningfully the temporal changes in agricultural landuse pattern, the investigator found noteworthy increase in percentage of cotton and hybrid jowar (17.11 per cent and 5.63 per cent) in the basin with the visible reduction of local jowar (20.00 per cent) during the study period (1975-76 to 1985-86) in the area under study.

7

Crop Combinations and Diversification

INTRODUCTION

After unfolding the agricultural landuse pattern in chapter 6, an attempt has been made to spotlight the spatial distribution of various crops, their growth and response to physio-socio-economic conditions prevalent in the basin. The cultivation of crops and their growth are closely related to the decision-making process on one hand and adoptation of innovation in agriculture, *i.e.* use of high yielding varities, improved and efficient implements, applications of chemical fertilizers and pesticides. Thus, the selection of crops for sowing in the field present composite picture of cultivation of crops in the region. The hectarage under individual crop gives relative strength and realistic picture of croplanduse in the analysis of crop ranking of the region. The ranks of crops and their combination provide spatial variation in the distribution patterns. In this respect the study of crop combinations and diverifications manifests the present agricultural scenario.

The following objectives have been put forth :

(*1*) To present areal strength of the crops grown in the basin by ranking and interpret the factors responsible for this rank distribution.

(*2*) To delineate the patterns of crop combination of the Yashoda basin by applying Doi's method and plot it in a regions of crop combination; and

(3) To identify the crop diversification patterns of the Yashoda basin by applying Bhatia's method of crop-diversification and find out the variables responsible for such patterns in the area under study.

The secondary data have been collected for various crops for the year 1985-86 in both kharif and rabi crops from Taluka headquarter office, Wardha District Gazetteer, Socio-economic Abstract of Wardha District and Census Handbook of Wardha District for 1981 are sources of data for this study. It is supplemented by numerous spot - inquiries. The area of crops have been converted into percentage (to net sown area) which are later on, used for ranking of crops to identify the relative strength of individual crop. The Doi's method has been applied to delineate the crop combination regions and crop-diversification has been computed to understand the magnitude of crop compitition among the crops grown in the basin.

CROP RANKING

The ranking obtained for twelve main crops in the basin to identify the relative significance of individual crop in cropping pattern. The first six ranking crops have been identified and mapped (Exhibits 7.2 and 7.3). The ranking crops and number of villages are shown in Table 7.1 and Exhibit 7.1.

First Ranking

Exhibit 7.2/A reveals the first ranking crops in the Yashoda basin. Three crops are identified occupying third rank, namely, cotton, local jowar and hybrid jowar. The above exhibit shows the dominance of areal strength of these three crops. Table 7.2 displays first ranking crops, villages and area in the study region. Cotton is major crop and it stands as first rank and is found to have largest coverage in the basin (Exhibit 7.2/A). The favourable environment in the region is responsible for growing cotton on fertile black soils. The

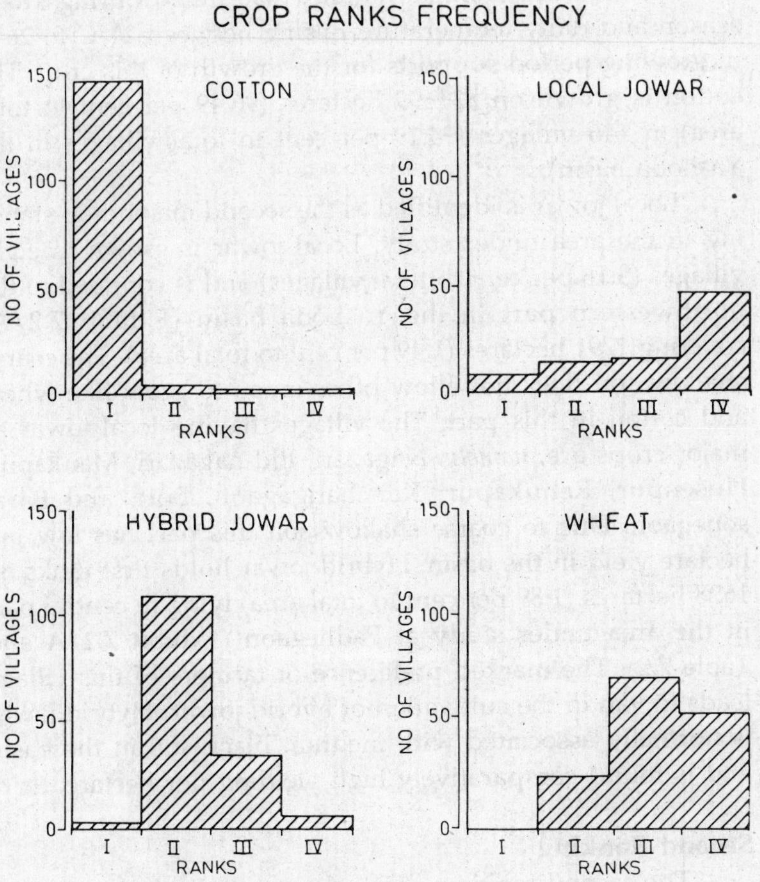
CROP RANKS FREQUENCY
COTTON
LOCAL JOWAR
HYBRID JOWAR
WHEAT
NO OF VILLAGES
RANKS
0
50
100
150
I
II
III
IV

EXHIBIT — 7.1

sufficient annual average rainfall (1050 mms) during kharif season and daily temperature raising between 21°C to 26°C in growing period supports for the growth of this crop. The cotton is grown on 831699 hectares (96.49 per cent to total area) in 146 villages (94.19 per cent to total villages in the Yashoda basin).

Local jowar is identified as the second major crop standing in the area under study. Local jowar is grown in eight villages (5.16 per cent to total villages) and is confined to the northwestern part in the Yashoda basin (Exhibit 7.2/A) covering 1291 hectares (1.49 per cent to total area). The coarse shallow soil does not allow other crops to grow like wheat and cotton in this part. The villages having local jowar as major crops are, *namely*, Nagazari, Bid-nagazari, Mankapur, Husenpur, Renukapur, Kanchangagaon, Bori, and Barasonegaon. Due to coarse shallow soil this part has low per hectare yield in the basin. Hybrid jowar holds first ranks on 1628 hectares (1.89 per cent to total area) lying in central part in the area under study at Padhegaon (Exhibit 7.2/A and Table 7.2). The marked preference of farmers of this village leads to van in the cultivation of hybrid jowar. Hybrid jowar is normally associated with medium black soil in the basin and is found comparatively high yielding per hectare.

Second Ranking

The second ranking crops present relatively large number of crops than those of first ranking. Exhibit 7.2/B reveals five crops ranked second. These five crops are cotton, local jowar, hybrid jowar, wheat and mug. Table 7.2/B displays second ranking crops, villages and area in the basin.

Hybrid jowar dominates the basin and holds the second rank in one hundred and twelve village (72.28 per cent to total villages) occupying wide-spread in size in the Yashoda basin. It covers 70409 hectare (81.78 per cent to total area) area in the basin (Exhibit 7.2/B). This crop is found in central and eastern parts in the basin and is associated with coarse

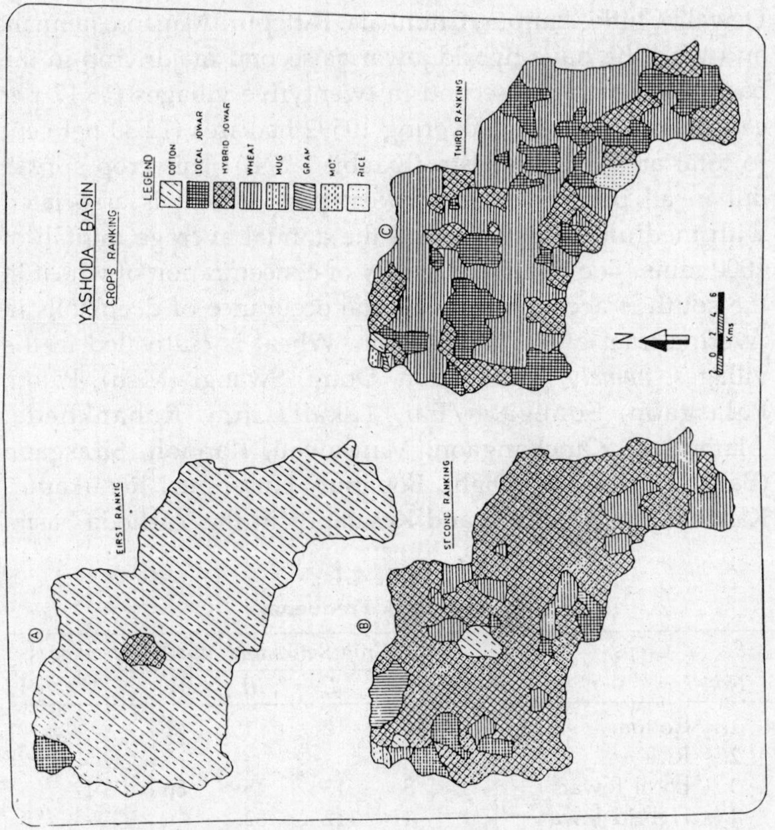
YASHODA BASIN
CROPS RANKING
LEGEND
COTTON
LOCAL JOWAR
HYBRID JOWAR
WHEAT
MUG
GRAM
MOT
RICE
A
FIRST RANKIG
B
SECOND RANKING
C
THIRD RANKING
N
0 2 4 6
Kms

EXHIBIT — 7.2

shallow soil to medium deep soil and annual average rainfall between 1000 mms and 1150 mms. Areal extent of hybrid jowar in the western part is restricted due to less amount of rainfall. The villages *namely*, Sirasgaon (Dnd), Waigaon, (Npn), Dewali, Chitki, Banpur, Chunhala, Fattepur, Ajagaon, Jamnala and Waddha have hybrid jowar as second major crop in the basin. Wheat ranks second in twentyfive villages (16.12 per cent to total villages) covering 10592 hectares (12.30 per cent to total area) in the basin (Exhibit 7.2/B). This crop spread out in all part except northwest part. Wheat is associated with medium black soils and the annual average rainfall of 1000 mms. The relative intensity of concentration of wheat in the south is accounted for by the occurance of deep soils in riverine tract of Yashaoda basin. Wheat is cultivated in the villages, *namely*, Muradgaon, Dotra, Swangi (Mgh), Paloti, Palasgaon, Sonegaon/Bai, Takali/Chnj, Rohankheda, Alamodah, Kanchangaon, Mansawali, Kanholi, Sirasgaon (Bazar), Takali/Jjr, Dighi, Bopapur, Husenpur, Renukapur, Kanchanpur, Nandora and Kesalapur in the Yashoda basin.

Table 7.1
Crops Rank Frequency

Sr. No.	*Crops*	*Crop Ranks and Number of villages* I	II	III	IV	V	VI
1.	Cotton	146	2	1	1	-	-
2.	Rice	-	-	-	1	63	28
3.	Local Jowar	8	13	45	86	1	-
4.	Hybrid Jowar	1	112	32	6	-	-
5.	Wheat	-	25	73	56	-	-
6.	Tur	-	-	-	-	24	35
7.	Linseed	-	-	-	-	3	17
8.	Mug	-	3	1	1	19	20
9.	Gram	-	-	1	2	8	12
10.	Sesamum	-	-	-	-	6	5
11.	Mot	-	-	2	1	4	26
12.	Groundnut	-	-	-	1	22	18
13.	Vegatable	-	-	-	-	5	3

Source : Author

Local jowar stands second in thirteen villages (8.38 per cent to total villages) covering 2838 hectates (13.29) per cent to total area) in the Yashoda basin. (Exhibit 7.2/B). The area of local jowar confined to central and western parts in the villages, *namely*, Laxminarayanpur, Khadaki, Lonsawali, Dorali, Dhamangaon, Bodad, Malkapur, Sekapur, Ajanavati, Adegaon, Raipur, Krishnapur and Meghapur have large coverage of local jowar among second ranking crops. The average annual rainfall in this part (1000 mms) seems to be adequate for the growth of local jowar.

Mug ranks second in three villages (1.93 per cent of total villages) covering 614 hectares (0.71 per cent to total area) in northwestern part in the area under study (Exhibit 7.2/B and Table 7.2). These villages are *namely*, Bori, Barasonegaon and mug is associated with coarse shallow soils in the basin as it appears to hardly crop to grow on even relatively less fertile soils. Cotton is identified as second ranking crop occupying 1635 hectare (1.89 per cent to total area) in two village located in the northwest part in the basin (Exhibit 7.2/B) and Table 7.2) *namely*, Barasonegaon and Padhegaon. These village farmers are preferred cotton as second crops after local jowar and hybrid jowar on coarse shallow soils.

Third Ranking

It is evident from Exhibit 7.2/C that the third ranking crops are widespread and patchy in their areal distributional pattern. This ranking crops rise seven in numbers. These seven crops are *namely*, cotton, hybrid jowar, local jowar, wheat, gram, mug and mot. Among these crops, wheat is the dominant crop in rank. Table 7.2/C displays the distribution of third ranking crops; villages and area in the basin. The outstanding feature of third ranking is the disappearance of cotton in basin except in village Nagazari. It is seen from Exhibit 7.2/C that wheat occupies dominant position covering 45862 hectares (53.27 per cent to total areas) in seventy three villages (47.10 per cent to total villages) and holds third rank in the Yashoda basin. Its areal extent sprawls over the north, central, and eastern parts in the study region. The wheat raising patch in the villages *namely*, Talegaon, Ajanavati, Ekurdi, Sonpeth, Allipur, Rohankheda, Gadegaon, Dapuri, Wallapur, Dewali, Selsura, Dagapur, Waddha, Babapur, Dogdoh, Palasgaon and Sarur and it is associated with medium black soil and average annual rainfall between 1000 mms and 1100 mms.

Table 7.2/A

Crops, Villages and Area in First Ranking

Crops	*Number of villages*	*Percentage of total villages*	*Area in hectares*	*Percentage of total area*
Cotton	146	94.19	83169	96.49
Local Jowar	8	5.16	1291	1.49
Hybrid Jowar	1	0.64	1628	1.89
Total	155	100.00	86088	100.00

Source : Author

Table 7.2/B

Crops, Villages and Area in Second Ranking

Crops	*Number of villages*	*Percentage of total villages*	*Area in hectares*	*Percentage of total area*
Cotton	2	1.29	1635	1.89
Local Jowar	13	8.28	2938	3.29
Hybrid Jowar	112	72.28	70409	81.78
Wheat	25	16.12	10592	12.30
Mug	3	9.93	614	0.71
Total	155	100.00	86088	100.00

Source : Author

Table 7.2/C

Crops, Villages and Area in Third Ranking

Crops	*Number of villages*	*Percentage of total villages*	*Area in hectares*	*Percentage of total area*
Cotton	1	0.64	476	0.55
Hybrid Jowar	32	20.64	18959	22.02
Local Jowar	45	22.05	20448	23.75

Wheat	73	47.10	45862	53.27
Mug	1	0.64	145	0.16
Gram	1	0.64	67	0.07
Mot	2	0.29	131	0.15
Total	155	100.00	86088	100.00

Source : Author

Local jowar holds third ranks and comprises fourtyfive villages (29.30 per cent to total villages covering 20448 hectares (23.75 per cent to total area (Exhibit 7.2/C). Local jowar is widespread in the basin and accounts for by traditional preference of farmers to cultivate this crop. The pockets of local jowar lying in western, central and eastern parts in the area under study. Local jowar in the west is associated with coarse shallow soils and average annual rainfall of 1000 mms. This western pocket includes the villages, *namely*, Sonegaon (Abj), Yesgaon, Muradgaon, Kelapur, Jamani, Loni, Dhonapur, Timapur and Akoli, while central and eastern pockets include villages namely, Padhegao, Ajanavati, Chikhali, Takali (Chnj), Jaitapur, Sarur, Aloda, Bhawanpur, Bhojankhea, Mund, Nimsada, Mendukdoh, Rohankheda, Pawani, Sirasgaon (Bazar), Nidha, Yerangaon, Ajagaon, Fattepur, Waigaon (Npn.), Taharpur, Kesalapur, Neri, Siwapur, Ashta, Jawalgaon, Sultanpur, Bhaiyapur, Shampur and Chitki. Local jowar here is cultivated on medium black soils.

Hybrid jowar stands as third ranking crop in the area under study. It covers 18959 hectares (22.02 per cent to total area) in thirty two villages (14.19 per cent to total villages (Exhibit 7.2/C and Table 7.2) in the basin. The cultivation of hybrid jowar indicates slight change to cultivate local jowar because it is high yield per hectare. Hybrid jowar can grow on coarse shallow soils. The villages lying in the third rank are namely, Kanchanpur, Bara-Sonegaon, Lonsawali, Dorali, Dhamangaon, Pimpri, Kurla, Sekapur, Malkapur, Muradgaon, Nandora, Akpal, Dotra, Sawangi (Mgh), Paloti Bopapur,

Sonegaon (Bai), Adegaon, Krishnapur, Pathari, Takacali (Drn.) Alamdoh, Kanchangaon, Mansawali, Kanholi, Takali (Jjr) and Nagthana.

Mot holds third rank in two villages covering 131 hectares area (0.15% to total area) in the Yashoda basin (Exhibit 7.2/C and Table 7.2/C). These villages are Nagazari and Bidnagazari. Hot tolerates shallow soils in the basin. Cotton, mug and gram stand as the rank in the area under study (Exhibit 7.2/C) and Table 7.2). These three crops cover 8.88 hectare (2.52 of total area) and these crops have a relatively law yield per hectare as are gram in shallow soil. Cotton is grown at Bori, (478) Mug at Renukaur (145 hectares) and gram at Mankapur (87 hectare). These villages are located in northwest part in basin.

Fourth Ranking

Fourth ranking crops present a much varied distribution both in space and the number of crops involved (Exhibit 7.3/A and Table 7.3/A). There are eight crops *namely*, wheat, local jowar, hybrid jowar, mug, gram rice, mot and groundnut (Exhibit 7.3/A and Table 7.3/A). Local jowar holds largest coverage among fourth ranking crops in the Yashoda basin. It is cultivated in central, east and southern parts in eighty villages (51.61 per cent to total villages) in the area under study and it covers 45516 hectares (52.87 per cent to total area, Exhibit 7.3/A). The major villages having dominance of local jowar are *namely*, Agargaon, Timapur, Akoli, Palasgaon, Dewali, Waifad, Padhegaon, Chitki, Jamatha, Allipur, Mansawali, Nidha and Sirasgaon (Bazar) wheat is followed by local jowar stands in fourth rank in the basin covering 32130 hectares (37.32 per cent to total area). Wheat cultivation has found patchy distribution in western, central and northern parts in the basin.

Hybrid jowar ranks fourth and is confined to five villages (3.22 per cent to total villages) in the western and eastern parts covering 7066 hectares (8.21 per cent to total area). Exhibit 7.3/A shows the areal distribution of hybrid jowar in the basin. The western part consisting villages *namely*, Laxminarayanpur, Bodad, Khadki, Muradgaon and in the eastern part at Rohankheda. This suggests marked preference of hybrid jowar by farmers. Hybrid jowar is associated with medium black soils. Gram is grown in two villages, *namely*, Bid-Nagazari and Kanchanpur and is confined to northwest part covering 64 hectares (0.07 per cent to total area) in the basin in coarse shallow soils. Rice, mot, mug and groundnut are grown in one village each. These villages are *namely*, Sawangi (Mgh), Mankapur, Husenpur, and Nagazari, respectively in northwest part in the Yashoda basin covering 1310 hectares (1.50 per cent to total area) on coarse shallow soils resulted low yield per hectare.

Table 7.3/A
Crops, Villages and Area in Fourth Ranking

Crops	*Number of villages*	*Percentage of total villages*	*Area in hectares*	*Percentage of total area*
Wheat	58	37.41	32130	37.32
Local Jowar	86	55.48	45581	52.87
Hybrid Jowar	5	3.22	7068	8.21
Gram	2	1.29	64	0.07
Rice	1	0.64	959	1.11
Mot	1	0.64	145	0.16
Mug	1	0.64	139	0.16
Groundnut	1	0.67	67	0.07
Total	155	100.00	86088	100.00

Source : Author

Fifth Ranking

The number of crops in fifth rank is ten and are spread throughout the basin. The crop-distributional pattern in this ranking is more fragmented and diversified (Exhibit 7.3/B). There is increased crops ten in number that include rice, tur, mug, gram, sesamum, local jowar, mot, groundnut, linseed and vegetables. Rice, tur and groundnut are dominant crops

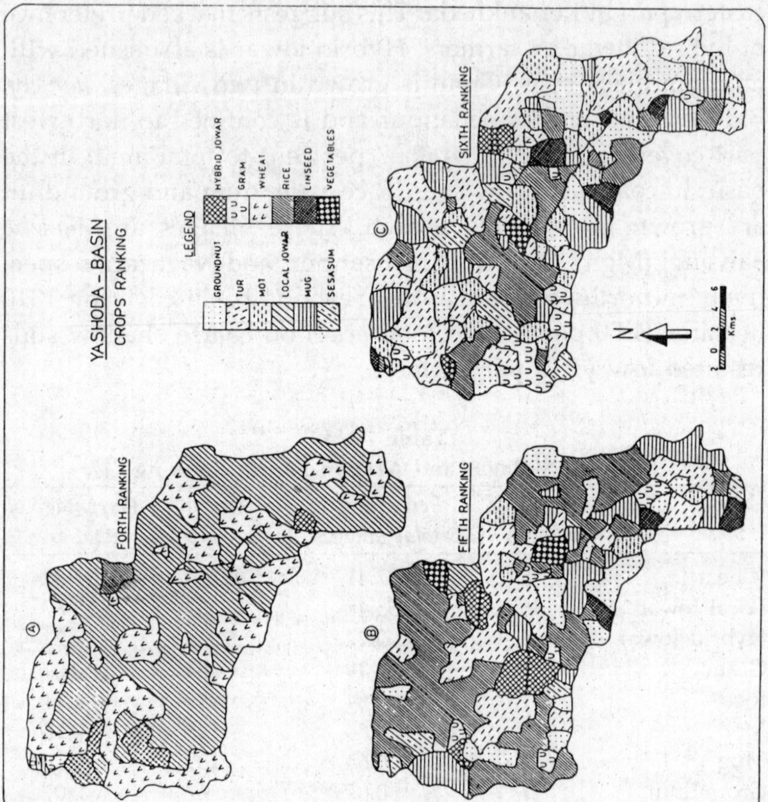
YASHODA - BASIN
CROPS RANKING
LEGEND
GROUNDNUT
TUR
MOT
LOCAL JOWAR
MUG
SESASUM
HYBRID JOWAR
GRAM
WHEAT
RICE
LINSEED
VEGETABLES
FORTH RANKING
FIFTH RANKING
SIXTH RANKING
A
B
C
N
0 2 4 6
Kms

EXHIBIT — 7.3

in the basin. (Table 7.3). Rice covers 35729 hectares (41.50 per cent to total area) comprising sixtythree villages (40.64 per cent of total villages) in the northern and eastern parts in the basin. The largest hectarage of rice is found in the villages, *namely*, waifad, Lonsawali, Tigaon, Dotra, Dhotra, Degaon/ Mskn, Yesgaon, Sonegaon (Abj). Allipur, Ekurli and Sonpeth. Tur is mostly grown in the central part covering 16554 hectares (19.22 per cent to total area) on medium black soils (Exhibit 7.3/B).

Table 7.3/B

Crops, Villages and Area in Fifth Ranking

Crops	*Number of villages*	*Percentage of total villages*	*Area in hectares*	*Percentage of total area*
Rice	63	30.64	35729	41.50
Tur	29	18.70	16554	19.22
Linseed	3	1.93	1196	1.38
Groundnut	21	13.54	9740	11.31
Hug	17	10.96	8225	9.55
Gram	7	4.51	2939	3.41
Seasamum	5	3.22	2613	3.03
Local Jowar	1	0.64	959	1.11
Mot	4	2.58	1872	2.17
Vegetable	5	3.22	6261	7.27
Total	155	100.00	88088	100.00

Source : Author

Groundnut cultivation is found in the southern and eastern parts in patchy forms in 9740 hectares (11.31 per cent to total area) in seventeen villages. Gram is grown on 2939 hectares (3.41 per cent to total area) comprising villages, namely, Malkapur, Adegaon, Fattepur (Rt.), Timapur, Alamdoh, Ajanavati and Akoli, Sesamum and vegetable hold fifth rank in five villages each covering 2013 hectares and 6265 hectares respectively. Sesamum is cultivated in the villages, *namely*, Husenpur, Renukapur, Kurzadi, Takali and Isapur, while vegetable is grown, in Sindi/Mgh., Waigaon/

Npn, Dewali, Selukate and Salod. Mot is cultivated in 1872 hectares on coarse shallow soils in the villages, namely, Bhagwa on, Pathari and Banpur. Local jowar stands as fifth ranks in the basin covering 959 hectares area. It is grown at Sawani (Mgh). Linseed is grown in 1198 hectares (1.38 per cent of total area) on medium to deep black soils in the villages of Konhali, Krishnapur and Rohankheda in the Yashoda basin.

Table 7.3/C
Crops, villages and Area in Sixth Ranking

Crops	*Number of villages*	*Percentage of total villages*	*Area in hectares*	*Percentage of total area*
Rice	24	15.48	12929	15.01
Tur	47	30.32	25582	29.71
Mug	17	10.96	12045	13.99
Mot	24	15.48	11666	13.55
Gram	12	7.74	5037	5.86
Linseed	6	3.87	3166	3.67
Sesamum	5	3.22	3716	4.31
Groundnut	16	10.32	9768	11.34
Vegetables	3	1.93	2042	2.37
Wheat	1	0.64	137	0.15
Total	155	100.00	86088	100.00

Source : Author

Sixth Ranking

There are ten crops which emerge as sixth in rank in the Yashoda basin. These crops are, rice, tur, mug, mot, gram, sesamum, groundnut, wheat, linseed and vegetables. Sixth rank crops indicate diversity in number and crop-distribution in the area under study (Exhibit 7.3/C and Table 7.3/C). Among the ten crops, tur is principle crop found largest coverage on 25582 hectares (29.7 per cent to total area) in northern and eastern parts in 47 villages (30.32 per cent total villages). The largest coverage of tur is found in the villages *namely*, Waifad, Laxinarayanpur, Neri, Chitki, Dhotra, Umari,

Sawangi/Mgh., Salod (Hpr.), Bbugaon, Ashta, and Allipur. Rice and mot are cultivated in twentyfour villages under each crop covering 12929 hectares (14.88 per cent to total area) and 11666 hectares (13.55 per cent to total area) respectively in the central part in the basin in the villages *namely*, Dewali, Jamani, Waddha, Dagapur and Narayanpur while western parts rice is cultivated in the villages *namely*, Timapur, Kurla, Sekapur and Nandpur. Rice is associated with coarse shallow soils. Groundnut is cultivated in sixteen villages (38.70 per cent of total villages) covering 9768 hectares (11.34 per cent to total area) in the western part, while eastern part groundnut is grown in the villages *namely*, Kurzadi, Karla, Degaon (stn.), Dighi, Sonegaon (Bai), Mug is found in numerous pockets in fragmented form comprising seventeen villages (10.96 per cent to total villages) in the central and western parts in the Yashoda basin. Gram is associated with medium black soils on 5037 hectares (5.85 per cent to total area) (Exhibit 7.3/C and Table 7.3/C). Linseed and sesamum area, *namely*, Kanchanpur, Bara-Sonegaon, Pimpri, Loni, Agargaon, Nagazari, Isapur, Muradgaon, Jamnala, Banpur, Bhagawa, Mankapur. The vegetable is dominant in three villages covering 2042 hectares (2.37 per cent to total area) in Selukate, Selsura and Kawatha. Wheat is associated with medium black soils at Taharpur covering 139 hectares in the Yashoda basin.

CROP COMBINATION

Recently, the crop combination analysis in geographical studies has gained momentum and its importance is increasing day by day. Any study of crops on regional scale must take into consideration the combinational analysis and the relative position of crops. Such analysis would ultimately minimise the change of oversimplified generalization (Ali, 1978). Combination studies are fruitful in many ways, firstly, they provide an adequate understanding of an individual crop geography. Secondly, combination is in itself an integrative reality that demand definition and distribution analysis, and lastly crop combination regions are essential for the construction of still more complex structure of vivid agricultural region (Weaver, 1954).

Combination Technique

A number of statistical technique have so far been introduced by geographers agriculturists, and economists to demarcate crop combination regions. The introduction of these crop-combination methods by Weaver (1954), Thomas (1963), Coppack (1964), Johnson (1958), Rafullah (1956), Bhatia (1960), Athawale (1966), Ayyar (1909), and Doi (1959). But in the present study Doi's technique (1959) has been specially used. It is modified form of Weavers' technique (1954) and is considered to be the easiest type of combination analysis. It is known as minimum deviation method.

Doi's Crop Combination Method

Kikukazu Doi introduced new technique of crop combination in 1959. This method substitutes the standard deviation $Ed\frac{n}{2}$ with the some of squared difference Ed^2. And the combination having the smallest Ed^2 is actually found by consulting a one sheet table whose requires the summary up of percentages instead of squaring the inferences.

In brief, instead of squaring the differences, it is required to sum up the percentage and then consult the table for the critical value of the next element at that accumulated percentage level. If the critical value is higher than the actual percentage, the crop is not considered, but if otherwise, the value is lower than the crop percentage, the crop is included in the combination.

To illustrate the statistical procedure involved the case of Kurzadi village in Yashoda basin may be cited. Individual crop percentage at kurzadi village during 1985-88 are given below Cotton = 42.09, Hybrid Jowar = 18.49, Local Jowar = 8.02, Tur = 5.43 and Gram = 5.08.

Table 7.4

Crop Combination (After Doi)

					Rank of Elements					
0	*1*	*2*	*3*	*4*	*5*	*6*	*7*	*8*	*9*	*10*
					Critical Values					
Cumulative	90							8.84	7.60	6.67
Percentage	85				12.93	10.00	8.17	6.91	5.99	5.29
of higher	80			13.83	10.00	7.85	6.46	5.49	4.78	4.23
ranking	75		16.67	10.57	7.75	6.13	5.06	4.32	3.76	3.33
elements	70	27.64	12.25	7.93	5.96	4.65	3.85	3.29	2.87	2.55
	68	23.54	10.73	6.98	5.17	4.11	3.11	2.85	2.43	2.15
	65	18.38	8.66	5.63	4.19	3.34	2.77	2.37	2.07	1.84
	60	11.27	5.46	3.59	2.68	2.14	1.78	1.52	1.33	1.18
	55	5.38	2.68	1.73	1.29	1.04	0.86	0.74	0.64	0.57
	50	0.00	0.00	0.00	0.00	0.00	0.00	0.00	0.00	0.00

Source : Doi's "The Industrial structural of Japanese Prefectures." Tokyo proceedings of the International Geographical Union, Section on Human Geography. Regional conference in Japan, 1957, pp. 310-316.

The critical values for only those sums of percentages which are required in the present case, all those crops are included in the combination whose accumulated percentage is less than 50, as the critical value for all the elements at 50 is 0.00 (Critical value Table 7.4). At Kurzadi the first crop occupies 42.09 per cent so that the next crop is automatically included. It makes the sum of the two crops (42.09 + 18.49) = 60.58. Now under the heading. 60, 5.46 is the critical value for the third ranking crop. Since third crop occupies 8.02 per cent area is larger than 2.60, it is included in the combination. The accumulated percentage of three elements (42.09 = 18.49 + 8.02) = 68.60, where the critical value for the fourth element is 6.98. Thus the third crop is also included. The sum of percentages of these four crops is (42.09 + 18.49 + 8.02 + 5.43) = 74.03 and corresponding critical value for 74.03 (or 75) is 7.75. As tur, with 5.43 per cent is smaller than this critical value. Tur and gram crops are excluded from the combination. Thus three crop combination (cotton + hybrid jowar + local jowar) is alloted to Kurzadi village for 1985-1986.

Crop Combination — Application and Results

The five crop-combinations have been identified for Yashoda basin (Exhibit 7.4). The villages and area under each crop combination are shown in Table 7.4.

Monoculture

Local jowar is identified as monoculture occupying only 0.32 per cent area in the Yashoda basin, it is occurred in two villages *namely*, Bid-nagazari and Mankapur in the northwest part of coarse shallow soils where rainfall receiving 900 to 950 mms (average annual rainfall). The monoculture dominant in the aforesaid area, has low yield per hectare and does not add much to the output of this crop (Exhibit 7.5/A and Table 7.4).

Table 7.4
Crop Combinations in Yashoda Basin

Combinations types	*Crops in combination*	*Number of villages*	*Area in hectares*	*Percentage of total Area*
Monoculture	LJ	3	276	0.32
Two crop combination	LJ/W/HJ/C	12	5338	6.43
Three crop combination	LJ/HJ/WC/R/C	86	50743	56.07
Four crop combination	LJ/HJ/W/C/R/V	53	299332	34.07
Five crop	LJ/W/G/HO/SE	1	199	6.23

Source : Author

Note :
LJ = Local Jowar; HJ = Hybrid Jowar
W = Wheat; C = Cotton
R = Rice; MO = Mot,
SE = Sesamum; G = Gram
V = Vegetables.

Two Crop Combination

Four crops, namely, (cotton, hybrid, local jowar and wheat enter into two crop combination in different villages. Exhibit 7.5/B reveals that the two crop combination area is visibly widespread in the central and north-western parts in

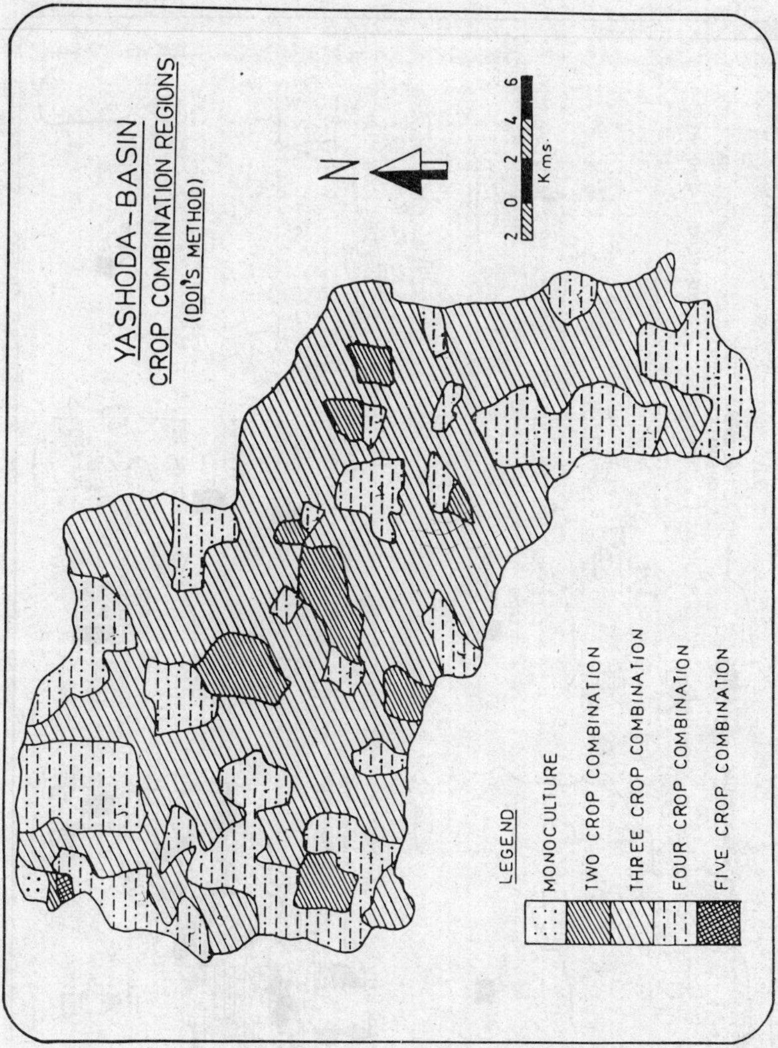
YASHODA-BASIN
CROP COMBINATION REGIONS
(DOI'S METHOD)
N
2 0 2 4 6
Kms.
LEGEND
MONOCULTURE
TWO CROP COMBINATION
THREE CROP COMBINATION
FOUR CROP COMBINATION
FIVE CROP COMBINATION

EXHIBIT — 7.4

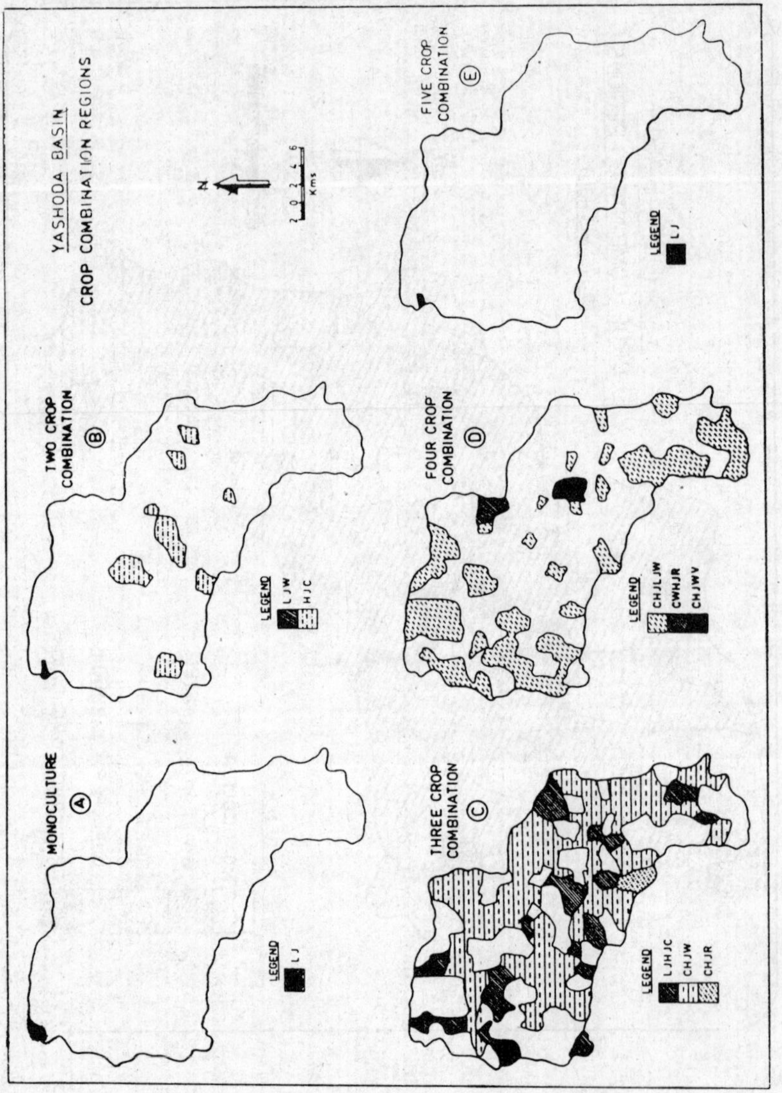
YASHODA BASIN
CROP COMBINATION REGIONS
N
2 0 2 4 6
Kms
MONOCULTURE
Ⓐ
LEGEND
LJ
TWO CROP COMBINATION
Ⓑ
LEGEND
LJW
HJC
THREE CROP COMBINATION
Ⓒ
LEGEND
LJHJC
CHJW
CHJR
FOUR CROP COMBINATION
Ⓓ
LEGEND
CHJLJW
CWHJR
CHJWV
FIVE CROP COMBINATION
Ⓔ
LEGEND
LJ

EXHIBIT — 7.5

the basin. Table 7.5 displays the two crop combination of crops in order, villages and area in the region under study. It is observed from Exhibit 7.5/B that largest area of two crop combination is in the central part covering 5399 hectares (6.24 per cent to total area) and lowest in the northwest part having 193 hectares (0.16 per cent to total area).

Table 7.5
Two Crop Combinations

Sr. No.	*Crops in order*	*Number of villages involved*	*Area in hectares*	*Percent of total area*
1.	Local Jowar/Wheat	1	139	0.16
2.	Hybrid Jowar/Cotton	11	5399	6.24

Source : Author

Cotton is the dominant crop in the central part and enters in combination with hybrid jowar in the basin. This combination appears in patchy form in the villages, *namely*, Padhegaon, Agargaon, Timapur, Takali, Muradgaon, Waddha, Banpur and Dattapur in the area of medium black soils. The eastern part in the basin has three patches of two crop combination area found in fragmented form where cotton associates with hybrid jowar local jowar is associated with wheat at Husenpur in northwest part and has two crop combination in the study region.

Three Crop Combination

Three are three and numerous small patches of this category in the Yashoda basin where three combination of various crops have been observed (Exhibit 7.5/C). Five crops enter in this combination are, *namely*, local jowar, hybrid jowar, cotton, wheat, and rice. Three crop combination is dominant on 50743 hectares (58.94 per cent to total area). This crop combination appears in the central and southern parts in the study area (Exhibit 7.5/C). Table 7.6 shows the crops in order, villages and area of three crop combinations in the basin. (Exhibit 7.5/C). It is obvious from Exhibit 7.5 that largest area of three crop combination of cotton, hybrid jowar and wheat extends over the central and eastern parts and spills over to southern part also. Medium black soils and average annual rainfall between 1000 mms to 1100 mms to seem to be favourable for this combination in the villages, *namely*, Umari, Dahagaon (Mskn.), Rotha, Degaon (Stn.), Neri, Sindhi (Mgh.), Padhegao, Narayanpur, Ashtrapur, Selsura, Chunhala, Aminpur, Wallapur, Asala, Rampur, Bhivapur, Dewangan, Ambapur, Talegaon, Sonpeth, Ekurli, Kawatha, Yesgaon, Nagazari, Belgaon, Digdoh, Dewali, Krishnapur, Chinchala, Mamdapur, Borgaon, Allipur, Sonegaon (Khr.), Pathari, Alamdoh, Kachangaon, Mansawali, Dhotra, Nagthana, Paloti, Nandora, Muradgaon, Wai, Dighi, Bopapur, Sonegaon (Bai), Babapur, Bodad, Kesalapur and Narayanpur. These fiftyseven villages (35.77 per cent to total villages) cover 56828 hectares (42.77) per cent to total area).

Table 7.6
Three Crop Combinations

Sr. No.	*Crops in order*	*Number of villages involved*	*Area in hectares*	*Percent of total area*
1.	Local Jowar/Hybrid jowar/cotton	26	12193	14.16
2.	Cotton/Hybrid jowar wheat	57	36838	42.77
3.	Cotton/Hybrid Jowar/Rice	1	1005	1.16
4.	Cotton/Local jowar jowar/wheat	2	717	0.82

Source : Author

The combination of cotton, hybrid jowar enter with rice is observed in south part in Giroli in 1005 hectares (1.16 per cent of total area) attributed to the presence of deep black soils in the basin. Three crop combination of cotton, local jowar and hybrid jowar is observed in small patches in fragmented in nature throughout the basin covering 12193 hectares (14.16 per cent to total area. Exhibit 7.5/C and Table 7.6 show this combination in the villages, *namely*, Bori, Dhamangaon, Meghapur, Sekapur, Akpal, Sindhi (kh.), Kurzadi, Kelapur, Afzalpur, Fattepur, Jawalgaon, Sirasgaon, Ajagaon, Taharpur, Sivapur, Asala, Bhawanpur, Karla, Jamani, Dhonapur, Ajanavati, Chikhali, Takali (Chnj.), Aloda, Pawani, and Bhaiyapur. These twentysix villages (16.66 per cent to total villages) are having three crop combination in which local jowar and hybrid jowar enter with the cotton in combination.

Cotton and hybrid jowar enter with wheat in this combination in two villages, *namely*, Bodad and Kesalapur in 717 hectares (0.82 per cent to total area) on black soils and hence has high yield per hectare.

Four Crop Combination

Four crop combination denotes the increase in the number of crops with comparative diversification in crop combination. Six crops in this combiantion entered are *namely*, cotton, hybrid jowar, local jowar, wheat, rice and vegetables. The region of four combination is wide spread in the western part in the basin (Exhibit 7.5/D, Table 7.7) displays the crop in order, villages and area in three crop combination.

It is observed from Exhibit 7.5/D that the area with four crop combination, comparatively, occupies less than that of the earlier crop-combination (29332 hectares). Furthermore, it is visible that two patches of four crop-combination appear in the basin. One at Waigaon having combination of cotton, hybrid jowar rice, and vegetable covering 1358 hectares (1.57 per cent to total area) and another at Sawangi (Mgh.) on 959 hectares (1.11 per cent to total area) appearing crops of cotton, hybrid jowar, local jowar and wheat in fiftyone villages (32.90 per cent to total villages). Exhibit 7.5/D shows four crop combination in the villages, *namely*, Kanchanpur, Barasonegaon, Laxminarayanpur, Chitki, Malkapur, Narayanpur, Pimpri, Khadki, Muradgaon, Sonegaon (Abj),

Loni, Akoli, Fattepur, Hirapur, Palasgaon, Nimgaon, Amboda, Lonsawali, Waifad, Wathoda, Amala, Tigaon, and Dorali in the basin.

Table 7.7
Four Crop Combinations

Sr. No.	*Crops in order*	*Number of villages involved*	*Area in hectares*	*Percent of total area*
1.	Cotton/Hybrid Jowar/ local jowar wheat	51	27015	31.34
2.	Cotton/Wheat/ Hybrid Jowar/rice	1	959	1.1
3.	Cotton/hybrid/ jowar/wheat/ vegetable	1	1358	1.57

Source : Author

A small patch of four crop combination (cotton, hybrid jowar, local jowar and wheat) appears in the south part covering 3560 hectares (4.13 per cent to total area) in the village, namely Nimasad, Takali (Drn), Mendukdoh, Mund, Rohankheda, Gadegaon (Rt.), Sirasgaon (Bazar), Kanholi, Takali (Jjr), Nidha and Yerangaon. This combination is attributed to medium black soils in the Yashoda basin. In the central part, combination of four crops (cotton, hybrid jowar, local jowar and wheat) are found in fourteen villages (9.03 per cent to total villages) namely, Bhaiyapur, Shampur, Dagapur, Thangddeo, Isapur, Adegaon, Raipur, Jaitapur, Sekapur, Jamnala, Kurzadi, Bhojankheda, Dapuri and Ajanavati in the basin covering 2580 hectares (2.99 per cent in total area).

Five Crop Combination

It can be safely inferred from Exhibit 7.5 that the entire picture of this combination is remarkably diversified not only in number of crops but it spatial disposition and distributional pattern are notably patchy. In the five crop combination local jowar and wheat is common to all. In addition, gram, mug, sesamum also enter in this combination. (Exhibit-7.5/E and Table 7.5). This combination is located exclusively at Renkapur in the northwest part on coarse shallow soils covering an extensive areal extent of 199 hectare (0.23 per cent to total area) in the Yashoda basin.

CROP DIVERSIFICATION

The crop-diversification technique is applied to compute crop-diversification pattern of the region. Its meaning is to raise variety of crops on arable land. It reflects the impact of physio-socio-economic variables. Moreover, it shows the contemporary competition among crops for an area, scope for rotation, the effect on double cropping, (Husain, 1979). The greater number of crops lead to greater competition, the higher is the magnitude of diversification.

Many geographers and economists so far have applied the concept of diversification in variety of sense. This concept, initially, was applied in the field of manufacturing to identify the degree of diversification and concentrations by Cleann (1930), later on, by Tree (1938) Florence (1942) and Rainwald (1949). Gibb Martin (1974) has used diversification concept in computing measurement of diversification of employment in industry. Among geographers, Bhatia (1965) adopted and introduced crop diversification technique in order to understand crop competition in the region followed by Jasbir Singh (1976), Ayyer (1969) modified Bhatia's method of crop diversification with accounting for those crops which occupy at least one per cent of the gross cropped area.

Crop Diversification Technique

In order to identify spatial pattern of crop diversification in present study, Bhatia's method has been adopted in modified form. The crops having five or less than five percentages have been excluded from computations. This modified formula expresses as:

$$\text{Index of Crop Diversification} = \frac{\text{Percent of Net Sown Area}}{\text{Number of 'n' crops}}$$

where — 'n' crops are those which individually occupy five or more than five per cent of crop to net sown area in the village.

Crop Diversification: Application and Results

The obtained results have been displayed in Exhibit 7.6 and Table 7.8. Table 7.9 shows crops in number, crops in competition, villages and area in crop diversification in the Yashoda basin.

Exhibit 7.6 portrays the spatial distributional pattern of crop diversification in the basin. Maximum crop diversification appears in northwestern part of Renukapur (9.25) and lowest at Nagazari (29). It is seen from the above Exhibit that four crop-diversification regions have been identified as :

(1) Area of high crop-diversification

(2) Area of moderate diversification

(3) Area of low diversification, and

(4) Area of very low diversification.

The four categories of crop diversifications, its class, magnitude, villages and area are shown in Table 7.8. It is observed from this table that the largest area appears in the moderate crop diversification class covering 59.51 per cent area in the basin and few villages characterised by very low crop diversification 0.5 per cent area in the study region) in the Yashoda basin. The four crops, *namely*, cotton, local jowar, hybrid jowar and wheat are in competition for diversification in one hundred and nineteen villages 76.79 per cent to total villages) in the Yashoda basin (Table 7.9).

Table 7.8
Patterns of Crop Diversification

Sr. No.	*Class of Crop diversification*	*Magnitude of diversification*	*Number of villages*	*Percentage of total villages*	*Area involved*	*Percent of area*
1.	0-10	High	1	0.64	199	0.23
2.	10-20	Moderate	80	56.61	51239	59.51
3.	20-30	Low	72	46.45	34519	40.09
4.	> 30	Very low	2	1.29	131	0.5

Source : Author

Table 7.9
Pattern of Crop Diversification

Crops in number	*Crops in competition for diversification*	*Number of villages*	*Percent of total villages*
One crop	Nil	Nil	0.00
Two crops	LJ + Mu	01	0.64
Three crops	LJ + Mu + Mo	01	0.64
Four crops	C+LJ+HJ+W	119	76.79
Five crops	LJ+W+MU+MOI+Gd	03	1.95
Six crops	R+C+G+HJ+V+L	29	18.70
Seven crops	LJ+W+Mu+Gd+T+Mo+C	01	0.64
Eight crops	C+R+LJ+HJ+G+T+R+S	01	0.64
		155	100.00

Source : Author

Note : C = Cotton, LJ = Local Jowar, HJ = Hybrid Jowar
W = Wheat, R = Rice, Mu = Mug
Mo = Mot, T = Tur, G = Gram
G = Gram, Gd = Groundnut, S = Sesasum
V = Vegetables. L = Linseed

Area of High Diversification

It is visible from Exhibit 7.6 that the area of high diversification appears in northwestern part at Renukapur (9.25) covering 199 hectares (0.23 per cent to total area) in the Yashoda basin. The largest number of crops are found in high degree of diversification. There are ten crop, namely, rice, local jowar, wheat, tur, linseed, mug, gram, sesamum, mot, groundnut, enter in this diversification. Uncertain and

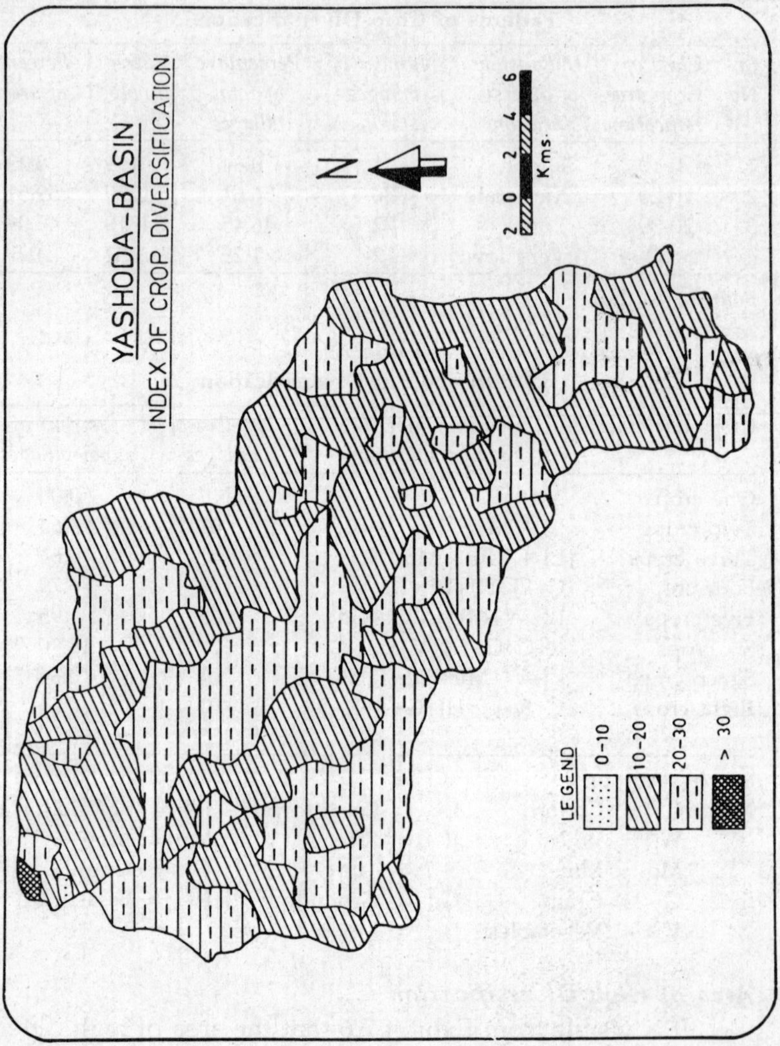
YASHODA-BASIN
INDEX OF CROP DIVERSIFICATION
N
2 0 2 4 6
Kms.
LEGEND
0-10
10-20
20-30
> 30

EXHIBIT — 7.6

low average annual rainfall, inferior soils of upland does not give high yield per hectare. Hence high diversification of crop is obvious in this part in the basin. The most of crops are generally sown in kharif season in the basin.

Area of Moderate Crop Diversification

This category covers 59.61 per cent area under moderate crop diversification in the basin. The area of crop diversification appears in east and south parts in the basin. The major largest patch appears in the riverine tract of Bhadadi (Exhibit 7.6) where high diversification is visible. The crops in diversification are cotton, hybrid jowar, wheat and tur on medium to deep black soils. This soil gives high yield per hectare in the rainfall receiving 1000 mms to 1050 mms. Eighty villages have occurred in central and eastern parts *namely*, Barasonegaon, Kurzadi, Waifad, Lonsawali, Dhamangaon, Sindi, Degaon (Mskn.), Dotra, Umari, Sindi (mgh) Sowangi (mgh.), Solad, Malegaon, Paloti, Jamatha, Selucate, Bhugaon, Banpur, Kurzadi, Sonegaon (Stn.) Wallapur, Neri, Dapuri, Hiarpur, Dattapur, Talegaon, Ajanavati and Ekurli in the basin.

The southern part in the basin identifies 10 to 20 index of crop diversification on deep soils tract where rainfall ranging between 1000 mms (average annual rainfall) in the villages namely, Sarur, Borgaon, Bhajankheda, Babapur, Takali, (Drn), Mendukdoh, Rohankheda, Gadegaon, Chanaki, Bhagawa, Kachangaon, Sirasgaon (Bazar), Yarangaon, Mansawali and Takali (Jjr), in the Yasoda basin. A small patch of moderate crop diversification appears in west part in the low rainfall (less than 900 mms rainfall) in the basin that puts limit on the growth of crops. In this category the villages includes, namely, Malkapur, Kelapur, Nandpur, Muradgaon, Khadaki, Pimpri, Loni, Sonegaon (Abj.), Dewali and Palasgaon.

Area of Low Crop Diversification

The area of low crop diversification covers 34519 hectares (40.09 per cent to total area) in the area under study

(Exhibit 7.6 and Table 7.9). The low crop-diversification region appears disintegrated in numerous patches throughout the basin. The largest patch of this category is found in the southern parts in the basin. Cotton, hybrid jowar, wheat, sesamum, linseed enter in this diversification. The western patch confines with average annual rainfall between 900 and 1000 mms and medium black soils has resulted low crop diversification in this part. It appears in the villages, *namely*, Laxminarayanpur, Kawatha, Chitki, Kurla, Badod, Degaon (Stn.), Jamani, Chikani, Padhegaon, Selsura, Waddha, Dagapur, Narayanpur, Ajunavati, Sonegaon (Bai), Agargaon, Timapur, Akoli, Hirapur, Fattepur and Ekpal. The south and southeastern parts in the basin are found low diversification of crops in villages namely, Chunhala, Aminpur, Taharpur, Bhawanpur, Dewangaon, Aloda, Sonpeth, Pathari, Nimsada, Ashta, Sultanpur, Jawalgaon, Siwapur and Alamdoh, Pawani, Sonegaon (Knr), Gadegaon, Kanholi, Nidha, villages have 20 40 index of crop diversification. The northern part has low diversification of crop is observed in patchy form in the villages *namely*, Bori, Dorli, Wardha, Amala, Tigaon, Rotha, Nagthan and Pathegaon due to inferior soils in this tract of the Yashoda basin.

Area of Very Low Crop Diversification

The northwestern upland part in the basin comprising two villages, namely, Nagazari and Bidnagazari have found very low crop diversification covering 0.5 per cent area of total basin (Exhibit 7.6 and Table 7.8). Local jowar, mot and mug are in competition in this area. These crops are grown on coarse shallow soils hence are unprofitable.

RESUME

In order to attempt an exposition of agricultural landuse pattern of the basin, the village has been taken unit for study to throw light on crop combination and diversification. The areal strength of individual crop has been discussed by ranking. This falls into six descending order. Cotton stands as first in ranking. While hybrid jowar first in second ranking crops. Cotton ranked in the first in one hundred and fourtysix villages (94.19 per cent to total villages) consequently, occupied the predominant position within the basin (2638.95 hectares) Hybrid jowar has been observed first in second ranking in one hundred and twelve villages (72.25 per cent to total area). Other crops such as local jowar, gram, wheat, rice have been ranked below cotton and hybrid jowar in the descending order.

The application of Doi's method shows the realistic picture of crop combination in the Yashoda basin. It has been observed that three and four crop combinations cover the largest areal extent in combination. Out of five crops in total, five crops *namely*, local jowar, wheat, rice, hybrid jowar and cotton enter in three crop combination in 86 villages on 58743 hectares (58.94 per cent of the basin). Out of the six crops *namely*, local jowar, hybrid jowar, wheat, cotton, rice and vegetables, four crops enter in four crop combination in 53 villages on 29332 hectares (34 per cent). Other crop combinations are of little significance in the basin as it gives low yield per hectare. The crops are *namely*, local jowar, wheat, gram, mot, sesamum, and cotton crops are in combination in three villages, namely Nagazari, Bidnogazari and Mankapur in northwest part on shallow course soils in the basin.

In order to understand the competition among crops in the basin, the crop-diversification has been computed by applying Bhatia's formula. According to Bhatia's method, two crop diversification categories have been revealed in the basin. The result of crop diversification established relationship with physio socio-economic conditions. The largest area covers with low crop diversification in 72 villages followed by moderate crop diversification in 80 villages on 34519 hectares (40.09 per cent) and 51239 hectares (59.21 per cent) respectively. The moderate crop diversification is found on extensive patch in the riverine area of Bhadadi where crops, *namely*, cotton, hybrid jowar, wheat, and tur are in this diversification, whereas the low diversification of crops is dominant in the western and southern parts on medium to deep black soils and crops, namely, cotton, hybrid jowar, wheat, sesamum and linseed enter in this diversification. The high crop diversification covers 199 hectares (0.23 per cent) confined to Renukapur in northwest part on low fertile soils. The very low crop diversification has found in two villages in Nagazari and Bidnagazari where mot and mug enter in diversification, owing to dominance of coarse shallow soils, with low yield per hectare in the Yashoda basin.

8

Quantitative Evaluation of Landuse

INTRODUCTION

In the preceding chapters 5 and 6, the general landuse and agricultural landuse patterns have been thoroughly studied and examined and the cropping patterns discussed. This study is based on the data and information to uncover the association between landuse and physio-socio-economic variables in the area under review. This established association was evaluated by suitable quantitative methods for further strengthening the results. From this point of view, the investigator tried to evaluate relationship between the variables by applied quantitative techniques in the Yashoda basin.

The present chapter deals with the quantitative evaluation of agricultural landuse pattern of the Yashoda basin with the following objectives :

(*a*) To find out the correlationship between selected agricultural landuse variables and environmental variables by applying quantitative techniques, *namely*, correlation coefficient and multiple regression, and,

(*b*) To identify the unique variance among selected variables of agricultural land use and environmental variables and to search for common variance as a set of new variables in the Yashoda basin and interpret the reasons.

DATA SOURCES AND METHODOLOGY

The data have been collected from secondary sources for the Year 1985-86 in both kharif and rabi crops from taluka headquarters office which, later on, converted into per cent to net sown area. Fifteen variables were carefully selected to assess the relationship for correlation coefficient in the basin. The significant variables in correlation coefficient were used for finding multiple regression (Table 8.1). Factor analysis has been computed for four factors which are accounted for 85.3 per cent variance (Table 8.3). These four factors are, further, used for factor identification and labelling which were later on used for computing scores for mapping.

CORRELATION ANALYSIS

In order to establish relationship between landuse and environmental variables, the correlation coefficient has been computed. The fifteen variables have been carefully selected from the set of variables of landuse and socio-economic conditions for one hundred and fiftyfive villages in the Yashoda basin. These chosen variables are as below :

X_1 = Percentage of net sown area to total geographical area.
X_2 = Percentage of cotton to net sown area.
X_3 = Percentage of local jowar to net sown area.
X_4 = Percentage of hybrid jowar to net sown area.
X_5 = Percentage of wheat to net sown area.
X_6 = Percentage of tur to net sown area.
X_7 = Percentage of mug to net sown area.
X_8 = Percentage of other pulses to net sown area.
X_9 = Percentage of linseed to net sown area.
X_{10} = Percentage of sesamum to net sown area.
X_{11} = Population density per sq kms.
X_{12} = Distance from major stream.
X_{13} = Percentage of irrigation to net sown area.

X_{14} = Slope less than 30 and,
X_{15} = Agricultural density.

Note : The detail information of variables used in landuse analysis is given in Table VIII/I.

Result of Correlation Coefficient

The degree of correlation has been computed by applying quantitative technique, *namely,* correlation coefficient for 155 x 15 data matrix. Pearson product movement correlation was applied. Student 't' test was, later on adopted to ascertain the significance of 'r' value at 0.001 level. The obtained results are shown in Table 8.1/A. The relationship between variables is summerised as below :

Net Sown Area (Area under cultivation)

The net sown area exhibits positive correlation with cotton, tur, population density, wheat and irrigation. The strong positive correlation of net sown area has found with cotton (0.600) while moderate correlation is established between net sown area and wheat (0.230) followed by population density (0.290) and irrigation (0.281) at 0.001 level. The spatial distribution of net sown area in the basin shows negative correlation with mug, other pulses, sesamum, local jowar, slope and agricultural density. This fact explains that the increasing net sown area does not proportionally increase the area under mug, other pulses, local jowar and agricultural density. It hass observed that the strong negative correlation of net sown area is found with local jowar (-0.566) while moderate negative correlation has established with wheat (-0.163) mug (-0342), pulses (-0.315), slope (-0.389) and agricultural density (-0.200). This indicates that the increase in areal extent under net sown area does not proportionally increase area under other pulses, wheat and mug.

Area under Cotton

The areal extent of cotton in the basin is associated with hybrid jowar, wheat, irrigation, population density and agricultural density and therefore, cotton shows positive correlation with these crops. This fact explains that increasing area under cotton is depending on the increasing area under above mentioned crops, while, variables *namely*, local jowar, tur, mug, other pulses, and slope are showing negative correlation in the area under review. The moderate positive association of cotton is found with hybrid jowar (0.340) and wheat (0.308) while linseed, population density, and agricultural density have low positive relation with the cultivation of cotton. (Table 8.1/A). The strong negative correlation is found with local jowar (-0.870), mug (-0.637), and slope (-0.691). The area under cotton decreases with increasing area under these crops. Low negative correlation has established with population density (-0.128), tur (-0.013), while, moderate relationship has been found with other pulses (0.482) in the basin.

Area under Local Jowar

The distribution of area under local jowar is positively associated with mug, other pulses and slope. There is an increase area under local jowar, the area under these crops increase proportionally. Local jowar can grow in low rainfall while mug also grow in low rainfall. Hence local jowar has strong positive correlation with mug (0.755) while moderate correlation has found with other pulses. The slope (less than 30) shows significant correlation with local jowar (0.645). The distribution of local jowar does not associate with wheat, hybrid jowar, tur, linseed, population density, distance from major stream, irrigation and agricultural density. The negative correlation has established due to the farmer's tendency to adopt new varieties of seeds for producing high yield per hectare. Moreover, local jowar can grow in all soil types, whereas wheat and hybrid jowar grow on medium to deep black soils. Therefore, hybrid jowar and wheat do not show significant association with local jowar. Local jowar has strong negative correlation with hybrid jowar (-0.551) and wheat

(-0.478). The variables namely, tur, linseed, population density have negative association with local jowar. Moderate correlation has established with irrigation (-0.187) and linseed (-0.155).

Table 8.1
The Variables used in Landuse Analysis

Variable No.	*Variables*	*Symbols*	*Formula*
1.	% Net Sown Area	= NA =	$\frac{\text{Total Geographical Area}}{\text{N.S.A.}} \times 100$
2.	% Area under Cotton	= CO =	$\frac{\text{Cotton Crop Area}}{\text{N.S.A.}} \times 100$
3.	% Area under Local Jowar	= LJ =	$\frac{\text{L Jowar Coop Area}}{\text{N.S.A.}} \times 100$
4.	% Area under Hybrid Jowar	= HJ =	$\frac{\text{H. Jowar Coop Area}}{\text{N.S.A.}} \times 100$
5.	% Area under Wheat	= WT =	$\frac{\text{Wheat Crop Area}}{\text{N.S.A.}} \times 100$
6.	% Area under Tur	= TU =	$\frac{\text{Tur Crop Area}}{\text{N.S.A.}} \times 100$
7.	% Area under Mug	= Mug =	$\frac{\text{Mug Crop Area}}{\text{N.S.A.}} \times 100$
8.	% Area under other pulses	= PU =	$\frac{\text{Other pulses crops Area}}{\text{N.S.A.}} \times 100$
9.	% Area under Linseed	= LD =	$\frac{\text{Linseed crop Area}}{\text{N.S.A.}} \times 100$
10.	% Area under Seasmum	= SM =	$\frac{\text{Sesamum Crop Area}}{\text{N.S.A.}} \times 100$
11.	Population Density	= PD =	$\frac{\text{Population}}{\text{Area}} \times 100$
12.	% Area under irrigation	= IR =	$\frac{\text{Irrigated Area}}{\text{N.S.A.}} \times 100$
13.	Agricultural Density	= AD =	$\frac{\text{Rural Population}}{\text{N.S.A.}} \times 100$

Table 8.1/A : Correlation Matrix

Variables	*X1*	*X2*	*X3*	*X4*	*X5*	*X6*	*X7*	*X8*	*X9*	*X10*	*X11*	*X12*	*X13*	*X14*	*X15*
X1	1.000														
X2	0.600	1.000													
X3	-0.566	-0.870	1.000												
X4	-0.163	0.340	-0.551	1.000											
X5	0.230	0.308	-0.478	0.168	1.000										
X6	0.019	-0.013	-0.195	0.065	0.288	1.000									
X7	-0.342	-0.537	0.755	-0.609	-0.550	0.443	1.000								
X8	-0.315	-0.482	0.380	-0.457	-0.388	-0.062	0.489	1.000							
X9	0.177	0.199	-0.155	-0.199	-0.027	-0.090	0.020	0.109	1.000						
X10	0.034	-0.128	-0.003	-0.286	-0.130	-0.061	0.173	0.483	0.254	1.000					
X11	0.290	0.167	-0.202	0.031	0.055	0.120	-0.182	-0.147	0.114	-0.072	1.000				
X12	0.003	0.136	-0.220	0.330	0.116	0.305	-0.294	-0.240	-0.176	-0.127	0.048	1.000			
X13	0.281	0.175	-0.187	0.021	0.031	0.089	-0.138	0.099	0.041	0.008	0.571	0.292	1.000		
X14	-0.389	-0.691	0.645	-0.224	-0.246	-0.050	0.448	0.247	-0.121	0.040	-0.139	0.000	-0.114	1.000	
X15	-0.200	0.107	-0.154	0.018	0.386	0.140	-0.148	-0.106	0.098	-0.051	0.954	O.074	0.566	0.107	1.000

(Value significant at 0.001 level)

Variables

X1 = Percent of NSA to TGS
X2 = Percent of Cotton to NSA
X3 = Percent of local jowar to NSA
X4 = Percent of hybrid jowar to NSA
X5 = Percent of wheat to NSA
X6 = Percent of tur to NSA
X7 = Percent of mung to NSA
X8 = Percent of other pulses to NSA
X9 = Percent of linseed to NSA
X10 = Percent of sesamum to NSA
X11 = Poulation Density per Sq. Kms.
X12 = Distance from major streams in kms.
X13 = Percent of irrigation to NSA
X14 = Slope less than 30, and
X15 = Agricultural Density.

Area under Hybrid Jowar

The area under hybrid jowar in the Yashoda basin shows positive correlation with net sown area, wheat and distance from major stream. While negative correlation with mug, other pulses, linseed, sesamum and slope. This fact can be explained that the cultivation of hybrid jowar show positive correlation (0.168). The area under hybrid jowar increases proportionally, with increasing area under hybrid jowar increases proportionally, with increasing area under wheat. This correlation suggests that hybrid jowar is concentrated in wheat dominating region. Hybrid jowar has established strong negative correlation with mug (-0.609) and other pulses (-0.321) while, moderate correlation with linseed (-0.199), sesamum (0.266) and slope (-0.224).

Area under Wheat

The spatial distribution of wheat is largely associated with population density, tur and agricultural density. The moderate correlation has established between wheat and agricultural density (0.366) and tur (0.288) while low positive relation is found between wheat and population density (0.116). This suggests that area under wheat increases with increasing area under tur. The variables namely mug, other pulses, sesamum and slope have negative correlation with the distribution of wheat. Wheat is mostly cultivated on medium to deep black soils where mug sesamum are extensively grown. Hence negative correlation has exhibited between wheat and mug and it shows strongly negative correlation (-0.550) while moderate negative correlation with other pulses (-0.388), and slope (-0.246). The low negative relation has found between slope and wheat (-0.130) in the basin.

MULTIPLE REGRESSION ANALYSIS

Multiple regressions have been computed to obtain the degree of correspondence between several variables in correlation coefficient. The variables found significant in correlation coefficient have been taken into account, here, to compute multiple regression. In the correlation coefficient analysis variables, *namely*, net sown area, cotton, local jowar, and hybrid jowar show significant correlation. Hence, these four variables were selected to compute the multiple equations for area under study. The following symbols are used in interpreting the result of multiple regression is given in (Table 8.2).

Table 8.2
Variables and Symbols used

Sr. No.	*Variables*	*Symbols used*
1.	Net sown area	NA
2.	Percent of cotton to net sown area	CO
3.	Percent of local jowar to net sown area	LJ
4.	Percent of mug to net sown area	MU
5.	Percent of other pulses to net sown area	OP
6.	Percent of sesamum to net sown area	SM
7.	Percent of irrigation net sown area	IR
8.	Population density	PD
9.	Slope	SL
10.	Percent of hybrid jowar	HJ
11.	Percent of wheat	WT
12.	Agricultural density	AD
13.	Distance form major stream	DS

Source : Author

Net Sown Area

The variable, net sown area has given a high multiple correlation value of 0.6337 for the area under study. In this regression set variables like cotton, local jowar, mug, other pulses, sesamum, population density, irrigation, and slope are included. The equation for net sown area is as follows :

NA = 64.7473 + 0.3519 CO - 0.2382 LJ + 0.8549 MU - 0.3301

OP + 0.2124 SM + 0.0087 PD + 0.1821 IR + 0.0.9096 SL

R^2 = 0.4016

Se = 5.54

R = 0.6337

The above equation and 't' statistics reveals that the area under local jowar and other pulses largely depends on net sown area. And net sown area is associated with cotton, mug, sesamum, population density and irrigation.

Area Under Cotton

This variable as an index of area under cotton shows spatial variations. Multiple regression analysis shows that variables like per cent of Hybrid Jowar (HP), per cent of wheat (WT), per cent of mug (MU) per cent of other pulses (OP), and degree of slope (SL) have governed the area under cotton and only net sown area shows a high degree of association with cotton. This crop has given high multiple correlation value of 0.8464 for the study area.

The equation for area under cotton is as below :

CO = 29.0206 + 0.3438 NA = -0.1801 HJ
-0.3708 WT = -1.7767 MU -0.7848 OP
-6.7552 SL.

R^2 = 0.7163

R = 0.8464

Se = 4.492.

Area under Local Jowar

This variable has been considerably governed by cotton (CO), Wheat (WT), and other pulses (OP). The multiple correlation value of 0.9236 which shows high degree of association for the study area. The equation for area under local jowar is as below :

LJ = 49.3651 -0.8144 CO -0.5148 WT
+1.5479 MU -0.7665 OP + 0.848 SL

R^2 = 0.8532
R = 0.9236
Se = 3.8467.

Area under Hybrid Jowar

The variable area under hybrid jowar has given high multiple correlation value of 0.7653 for the study region. In this regression set, variables like net sown area (NA), local jowar (LJ) other pulses (OP), sesamum (SM), distance from major stream (SD) were included.

The equation for area under hybrid jowar is as below :

HJ = 55.8751 -0.0950 NA -04996 CO
-0.6212 LJ -0.7028 OP -1.1885 SM -6.1095 DS.
R^2 = 0.5851
R = 0.7653
Se = 3.2935.

The above equation and 't' statistics reveal that the area under NA, CO, LJ, OP, SM and DS is highly correlated with area under hybrid jowar.

FACTOR ANALYSIS

The geographers have recently given more stress on the application of quantative techniques in geography in order to obtain realistic results. The quantitative techniques are of many types. Factor Analysis is one of it. In the present study, Factor Analysis, quantative technique, has been employed to search for a common variance as a set of new variable. This technique helps to exclude unique variances and shared common variances with that of other variables in the analysis. Furthermore, it ruled out all information which are of little significance.

The Rotated Factor Analysis (Varimax) technique has applied to identify the groups of variables with shared common variances. One hundred and fiftyfive observations and fifteen variables (155 x 15) were computed (Table 8.2). The first four variables accounted for 85.30 per cent variances, while remaining eleven variables explain only 14.7 per cent variances. These eleven variables show little significance in landuse analysis. Hence, the first four variables, being significant and influensive have been considered as cut of point. These four variables, are further, computed to compute loadings and scores on individual factor.

Table 8.3

Percent of Explained Variance Accounted for By Each Factor

Sr. No.	*Factor*	*Percent of explained variance*	*Commulated Percentage*
1.	1	43.10	43.10
2.	2	20.60	63.70
3.	3	13.60	77.30
4.	4	8.00	85.30
5.	5	6.30	91.60
6.	6	3.80	95.40
7.	7	2.30	97.70
8.	8	1.40	99.20
9.	9	0.80	100.00
10.	10	0.00	100.00
11.	11	0.00	100.00
12.	12	0.00	100.00
13.	13	0.00	100.00
14.	14	0.00	100.00
15.	15	0.00	100.00

Note : Factor Analysis has computed in Geography Department, University of Pune, Pune.

Identification of Inter-correlated Groups of Variables

Table 8.3 presents the Factor loadings and estimated communalities for first four factors. The factor loadings uncover the relationship between the variables. It is obvious from Table 8.4 that first two factors are significant in the basin, accounts for 43.10 per cent and other two factors do not show influence, hence are accounting little explain variances *i.e.* 13.6 per cent and 8.00 per cent respectively. The first two factors (Factor I and II) are graphically plotted to show concentration of interrelated variables in the Yashoda basin (Exhibit 8.1). The three intercorrelated groups have been identified in the present study. Two groups have positive intercorrelated, while third is negative. First group of variables (A) consists of net sown area and cotton. The communalities of cotton and net sown area are 0.95 and 0.47 per cent respectively. This fact explains that net sown area increases with increasing area under cotton proportionally. The second group (B) includes variables, *namely*, agricultural density, population density and irrigation. These variables are highly intercorrelated with each other showing positive relationship. The communalities of population density (0.92 per cent) followed by agricultural density (0.91 per cent) and irrigation (0.44 per cent). This group of variables indicate more concentration than earlier group as landuse is related to population and irrigation. Third group (C) presents negative intercorrelation between the variables namely, other pulses, mug, slope, and local jowar. Moreover, this group is inversely inter-correlated with earlier two groups. Therefore the location of earlier two groups are exactly opposite to third (Exhibit 8.1).

Factor Identification and Labelling

Factor Analysis brings out the hidden factors from the set of variables or shared common variances of the variables 85.30 per cent variance has been considered as cut-off point in present study. Accordingly, four factors have identified which were, later on, named and labelled as under :

Factor I : This new factor traces about 43.10 per cent variances in the Yashoda basin. This factor has been pulled closer to cotton and it shared loading is 0.944587. In addition to this, cotton variable shows comparatively higher loading than other shared variables (Table 8.4). Therefore, this factor can be labelled as *Cotton Factor*.

Factor II : This factor accounts for 20.60 per cent of variances in the present study. This factor shows high loadings

Table 8.4

Rotated Factor Loadings (Various Rotation) and Estimated Communalities

Variables	Factor Loadings				Communalities
	Factor 1	*Factor 2*	*Factor 3*	*Factor 4*	
1. Precent of NSA to TGA	0.611925	0.230079	-0.001860	-0.021353	0.47979
2. Percent of Cotton to NSA	0.944587	0.067407	-0.154124	-0.015498	0.95497
3. Percent of L. Jowar to NSA	-0.912505	-0.075099	0.078515	-0.336829	0.96992
4. Percent of H. Jowar to NSA	0.359163	-0.051829	-0.676710	0.193468	0.87163
5. Percent of Wheat to NSA	0.364435	-0.040135	-0.039558	0.601362	0.73731
6. Percent of Tur NSA	-0.036045	0.112759	-0.019779	0.62592	0.48978
7. Percent of Mug to NSA	-0.627058	-0.063925	0.341400	-0.425755	0.73513
8. Percent of Other pulses to NSA	-0.391045	-0.087302	0.642008	-0.099782	0.69689
9. Percent of Linseed to NSA	0.234814	0.096432	0.396373	-0.128266	0.33014
10. Percent of Sesamus to NSA	0.003855	-0.053670	0.641256	0.022717	0.52483
11. Percent of density	0.129062	0.954758	-0.017943	0.029288	0.92455
12. Distance from major stream	0.023949	0.104305	0.315006	0.384612	0.34268
13. Percent of irrigation to NSA	0.121743	0.608665	-0.016656	0.099517	0.44112
14. Slope	-0.787910	-0.055603	-0.201891	-0.033732	0.51491
15. Agricultural density	0.056704	0.953371	-0.007647	0.066151	0.91974
eigen values	4.43	2.12	1.40	0.83	-
Percent of Explained variance	43.10	20.60	13.60	8.00	-
Cummulated Percentage	43.10	63.70	77.30	85.30	-

Source : Author

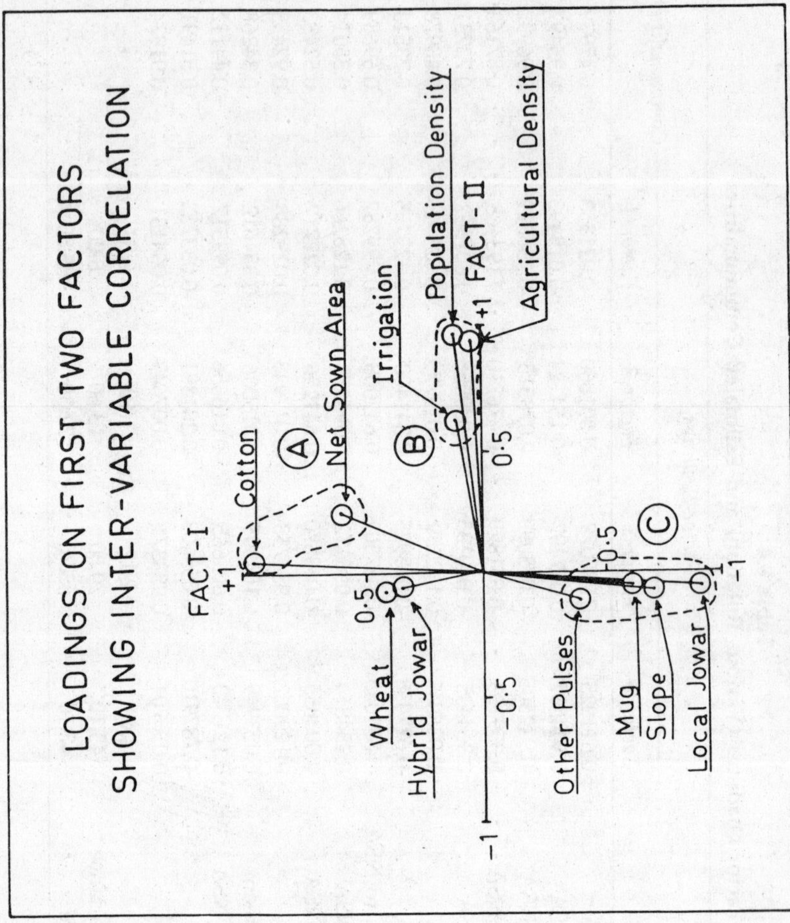
LOADINGS ON FIRST TWO FACTORS
SHOWING INTER-VARIABLE CORRELATION
FACT-I
FACT-II
Cotton
Net Sown Area
Irrigation
Population Density
Agricultural Density
Wheat
Hybrid Jowar
Other Pulses
Mug
Slope
Local Jowar
A
B
C
+1
0·5
-0·5
-1

EXHIBIT — 8.1

on two variables, *namely* population density and agricultural density having 0.945758 and 0.95337 loadings respectively. It is obvious that these two variables traces relatively high loadings than other variables. Hence this factor can be labelled as *Density Factor*.

Factor III : This new factor traces 13.60 per cent of explained variances. This factor has been pulled closer to sesasum and other pulses, loadings is 0.641256 and 0.642008 respectively. This factor can be identified as *Pulses and Sesamum Factor*.

Factor IV : Factor IV accounts for 8.00 per cent of explained variances and is closer to wheat and tur. Tur and wheat variables have high loadings than other involved variables. Wheat has 0.601362 and tur has 0.625920 loadings. Hence this factor can be labelled as *Wheat and Tur Factor*.

Factor Scores Mapping

Factor scores on each Factor was obtained for 155 x 15 data matrix. It was, later on, mapped and interpreted. Exhibit 8.2 shows spatial distribution of first four factors.

Scores on *Factor-I* indicates the spatial distribution pattern in the Yashoda basin (Exhibit 8.2/A). It is clear from this exhibit that high positive loadings are observed in south part in the study region where both cotton and net sown area are significantly high. In this part cotton acreage increases proportionally with increasing area under net sown area. North and west parts in the region has less scores. It explains that cotton and net sown area do not increase proportionally. Central part in the basin along river net sown area and cotton is found dominant.

Scores on *Factor-II* presents slightly different picture of spatial pattern than earlier. (Exhibit 8.2/B). In this spatial pattern positive loadings occur on irrigation, population density and agricultural density. High loading are found in east, north and west parts (more than 1 score) central part has high loading in the basin due to dominant of cotton and wheat, whereas negative loadings occurred on local jowar, hybrid jowar, mug, other pulses, sesamum and slope. The negative scores (0 to 1), are found in central and south (Exhibit 8.2/C) in the basin. Here area under locals jowar hybrid jowar, other pulses, mug, sesamum do not significantly increase with increasing population density, agriculture density and irrigation.

Scores on the *Factor III* reveals a comparatively more complex pattern than earlier two (Exhibit 8.2/C). But pockets in Factor III are visible. In this factor, variables, *namely*, other pulses and sesamum have high positive loading in central, western and east parts in the basin. (0 to 1 score). The highest scores is observed at Ajanavati (1) in the east in the study area due to correlation of other pulses and sesamum. The negative scores has occurred on hybrid jowar. The negative score regions are found in central and west parts where this crop does not extend hectarage due to dominance of cotton.

The *Factor-IV* scores presents realistic distribution pattern of loadings than Factors II and III (Exhibit 8.2/D). The positive loadings are found on wheat and tur variables score (0 to 1 score) where tur and wheat are dominant. The major dominant pockets of negative scores are observed in south part where mug has relatively less hectarage. Hence mug variable shows negative loadings. Area under mug does not significantly increase with increasing area under wheat and tur. The south and southeast parts in the basin is found dominant for mug.

Agricultural Regions (Based on Scores on Factors - I & II)

There are variety of approaches to delimit agricultural regions. Many geographers have applied statistical methods to demarcate agricultural regions which are mainly based on crops and their hectarage in a unit. In present study, investigator is attempted to identify agricultural regions on the basis of factors contributing in landuse analysis. And therefore, quantitative technique, *namely* Factor Analysis has been

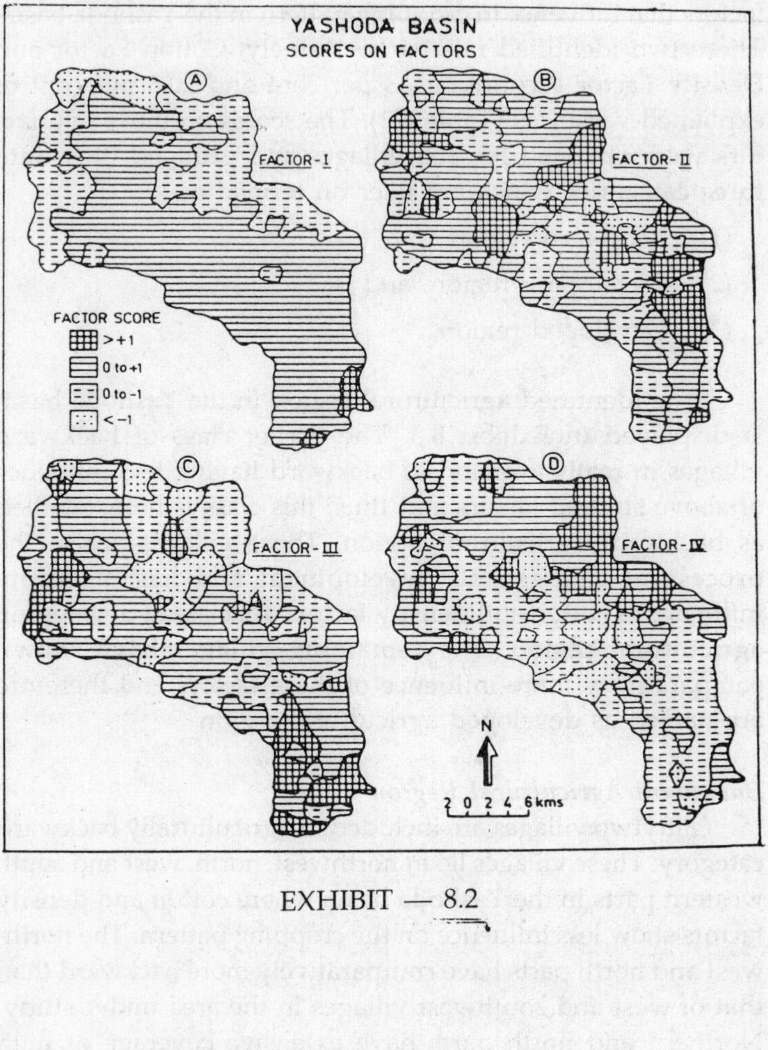
YASHODA BASIN
SCORES ON FACTORS
A
FACTOR - I
B
FACTOR - II
C
FACTOR - III
D
FACTOR - IV
FACTOR SCORE
> +1
0 to +1
0 to -1
< -1
N
2 0 2 4 6 kms

EXHIBIT — 8.2

adopted. The results of factor analysis (Table 8.4) unfolds two factors that influence in cropping pattern in the Yashoda basin. These two identified factors are namely, Cotton Factor and Density Factor accounting 43 per cent and 20.6 per cent of explained variances (Table 8.3). The scores on these two factors were obtained for 155 villages and grouped them into three categories which are later on termed as:

(1) Backward region

(2) Developing region, and

(3) Developed region.

The identified agricultural region in the Yashoda basin is displayed in Exhibit 8.3. The earlier class of backward villages in really a region of backward having less influence of above studied factors and thus, this class is here, labelled as backward agricultural region. Those villages are in the process of change and development and has moderate influence of cotton and density factors are classed developing agricultural region. The remaining villages have shown comparatively more influence of these factors and therefore are named as developed agricultural region.

Backward Agricultural Region

Fiftytwo villages are included in agriculturally backward category. These villages lie in northwest, north, west and south western parts in the Yashoda basin where cotton and density factors show less influence on the cropping pattern. The northwest and north parts have comparatively more backward than that of west and southwest villages in the area under study. Northern and north parts have extensive coverage of high relief, rugged and stony surface of Tigaon Hill and therefore those villages are located in north and northwest are found inferior soil and less possibility of expansion of net sown area. Net sown area in this part ranges between 40 to 60 per cent to total geographical area and barren land under net

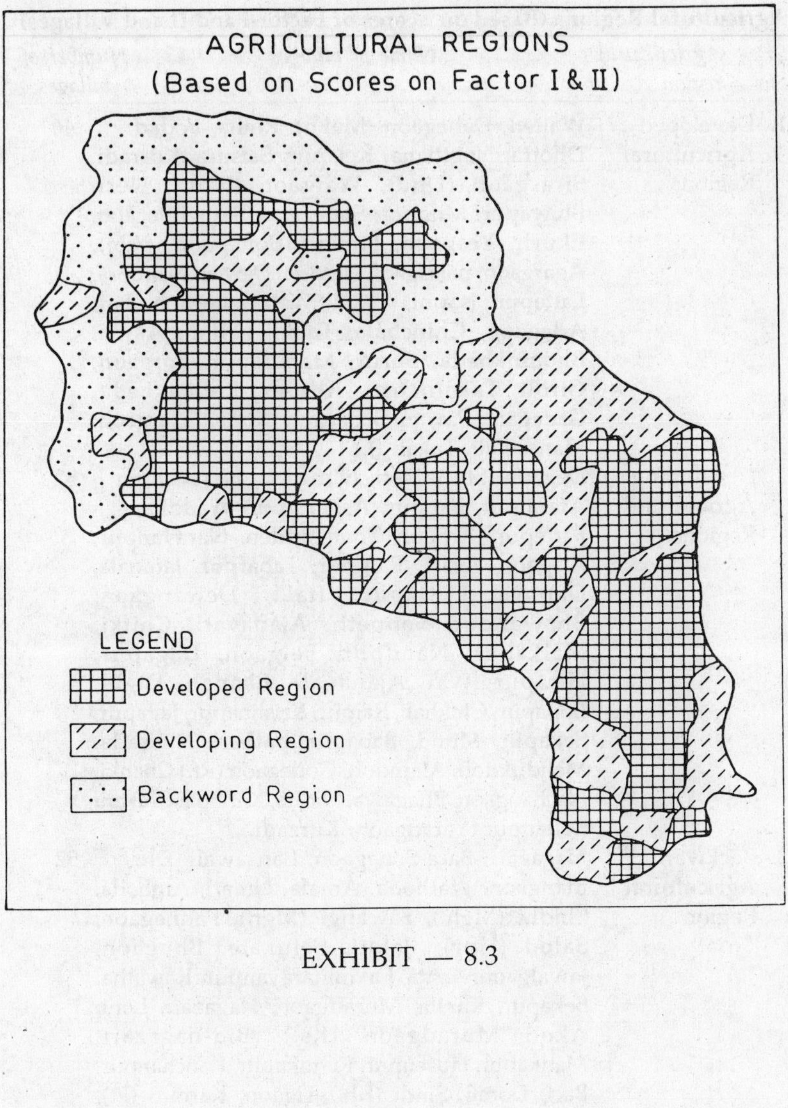
AGRICULTURAL REGIONS
(Based on Scores on Factor I & II)
LEGEND
Developed Region
Developing Region
Backword Region

EXHIBIT — 8.3

Table 8.4/A

Agricultural Regions (Based on Scores of Factor-I and II and Villages)

Sr. No.	*Agricultural region*	*Name of villages*	*Number of villages*
1.	Developed Agricultural Regions	Waited, Dahegaon (Makn), Rotha, Bodad, Dhotra, Nagthana, Kelapur, Selsura, Kurradi, Sirasgaon (Dnd), Waigaon (Npn), Neri, Bhiwapur, Mirapur, Sonegaon (Stn.) Talegaon, Ekurli, Yesgaon, Jamani, Sonegaon, (Abj), Agargaon, palasgaon Digdoh, Dewali, Nandora, Fattepur, Isapur, Takali(Mjr), Sonegaon (Bai) Adegaon, Chinchala, Takali (Chnj), Aloda, Bhojunkheda, Surur, Mamdapur, Borgaon, Giroli, Takali(Drn), Allipur, Rohankheda, Gadegaroi Pawani Sonegaon (Khnr), Sirasgaon, Mansawali, Takali (J.).	46
2.	Developing Agricultural Regions	Degaon (Stn.), Neri, Nimgaon, Narayanpur, Afaralpur, Banpur, Asharatpur, Waddha, Fattepur, Anipur, Thangaddeo, Narayanpur, Ajagaon, Asala, Rampur, Taharpur, Jamnala, Dapuri, Dattapur, Itala, Dewangaon, Bhawanpur, Sonpeth, Ajanavati, Chitki, Malkapur, Nandpur, Belgaon, Dagapur, Hirapur, Wai, Ajanavati, Ekapali, Dighi, Bopapur, Chikhali, Raipur, Krishnapur, Jaitapur, Sekapur, Mund, Babapur, pathari, Nimsada, Mendukdoh, Alamdoh, Godegaon (Rf.) Chanki, Nachangaon, Bhagawa, Nidha, Kanholi, Chikani Sultanpur, Yaradgaon, Kurzadi.	57
3.	Backward Agricultural Region	Nagazari, Bara-Sonegaon, Lansawali, Dhamangaon, Wathoda, Amala, Umari, Amboda, Sindhi (Mgh.), Sawangi (Mgh), Padhegaon, Salod (Hpr), Paloti, Selukate, Bhugaon, Jawalgaon, Ashta, Laxminarayanpur, Kawatha, Sekapur, Karala, Muradgaon, Nagazari, Loni, Akoli Muradgaon (Ks.), Bid-nagazari, Mankapur, Husenpur, Renukapur, Kanchanpur, Bari, Dorali, Sindi (Kh.) Tigaon, Karradi (Ft), Padhegaon, Bhaiyapur, Shampur, Hirapur, malegaon, Jamatha, Meghapur, Babpur, Kesalapur, Chunhala, Wallapur, Ambapur, Pimpri, Khadki, Timapur, Dhonapur (Rt).	52

Note : Scores has been computed in Geography Department, Pune.

available for cultivation is found between 10 to 20 per cent. Agriculturally this barren land has little use and therefore, cross like local jowar, mug, mot, sesamum are raised in this part. The soil of this region is coarse shallow containing 29.55 per cent fine sand and 22.20 per cent coarse sand. The cropping pattern due to inferior soil is largely influenced within backward agricultural region. The villages in this category are, *namely*, Nagazari, Bidnagazari, Lonsawali, Dhamangaon, Waifad, Amala, Umari, Amboda, Sidi (Kh.), Tigaon, Husenpur, Renukapur and Kanchanpur, local jowar and mug are dominant crops stands first and second rank in north and northwests parts. The agriculturally backward region extended in southwest, west and central parts (Exhibit 8.3) where net sown area is observed 50 to 60 per cent. It is comparatively more than northpart. The villages include in this category are Muradgaon, Laxminarayanpur, Kawatha, Kurla, Loni, Akoli, Khadaki, Timapur, Dhonspur, Pimpri and Sekapur in west, while Sindi (Mgh.), Sawangi (Mgh.), Bhugaon, Paloti, Selukate and Padhegaon n central part where cotton and density factor shows less impact than other two agricultural region. Among other crops in backward region are, cotton, hybrid jowar.

Developing Agricultural Region

Fiftyseven villages are found in the process of change and development and has moderate influence of cotton and density factors and therefore, this region is called as developing region. This developing region is noticed from Exhibit-8.3. It lies in central, east and south parts in the Yashoda basin between the tract of developed region. It is clear that developing region has some common characteristics of both backward and developed region. This region lies on medium black soil tract where percentage of cotton and hybrid jowar are increased after 1980 due to adaptation of new variety of seeds for sowing. Moreover, sprayings and application of chemical fertilizers are recently adopted in agricultural practice. The villages located in central part are really in the process of change in agricultural practice in order to get high yield per hectare. These villages are Wardha, Banpur, Ashtaratpur, Narayanpur, Asala, Rampur, Daput, Bopapur, Chikhali, Raipur, Krishnapur, Dighi, and Jaitapur. A small patch of agriculturally developing region is observed in south in the villages *namely*, Chanaki, Bhagawar, Yarangaon Nidha, Kanhali, while eastern villages are developing tract of agricultural region in which, Sultanpur, Ajagaon, Asharatpur, Banpur, and Narayanpur villages are included. The developing village have found in the stage of process of change and now farmers of these villages slowly realize the significance of agriculture in these villages. There is potentiality to raise larger hectarage under cotton and hybrid jowar by sowing improved varieties of seeds. Irrigation is also source of potentiality patch wel existed in these villages can be utilized for irrigating crops. Small size of holding is really a hindrance to make efficient landuse in developing region. In brief, with better planning and management the agriculture of this region may develop in near future.

Developed Agricultural Region

Fourtysix villages are included and is termed as developed agricultural region considering high positive scores on factors on cotton and density. Central and southern parts in study region have shown domination of these factors. This region is not as agriculturally developed region like other region of world but shows comparatively high developed than earlier two agricultural regions undertaken for present study. It is observed that these villages are located on medium to deep black soils tract and most of them have high percentage of net sown area to total geographical area. This region, physiographically lies in the Bhadadi-Punsoda doab and lower Yashoda river. Moreover, it is observed that southern villages have very high percentage of net sown area than central due to high fertility of soils. The villages included in lower Yashoda are namely, Sonegaon (Stn.), Aloda, Sarur, Chinchala, Takali (Chnj.), Borgaon, Giroli, Takali (Jjr), Allipur, Rohankheda, Godegaon, Pawani, Sonegaon (Khr.), Sirasgaon (Bazar), Mansawell, and Nidha. Fine grains transported by rivers are deposited in this low lying part hence soil is productive. The analysis of deep black soil in this region uncovers the percentage of clay. It found 43 per cent in this soil. Cotton stands first in rank and dominates the entire south part followed by hybrid jowar. Both crops have shown upward trend in all villages. Other crops sown in the region are mug, local jowar, mot, sesamum, groundnut but are found negligible hectarage. Kharif and rabi are two agricultural seasons among them kharif season is dominant as more than 70.9 per cent crops are raised in this period. Wheat occupies first position in rabi crops in this parts. Centrally located villages are also found to have high percentage of net sown area where cotton and hybrid jowar are predominant. It unfolds during interview that farmers in this tract are in a position to adopt new techniques and methods of agricultural practice. Improved seeds of cotton and hybrid jowar are used for sowing irrigation is found in the villages which are located on the river Yashoda. These villages are, Chanki, Mansawali, Yarangaon, Dewali, Sarur, Allipur, and Pawani. As a whole, this region is aptly termed as developed agriculture region in Yashoda basin.

RESUME

The presentation on quantitative evaluation of land use in the present chapter has successfully brought out relationship between landuse variables and environmental variables. The quantitative technique, *namely,* correlation coefficient, and multiple regression have employed. The correlation coefficient established positive correlationship between the variables. Cotton, hybrid jowar, and wheat have uncovered positive correlation with the net sown area. The net sown area established high positive correlation with cotton (r = 0.600) and moderate correlation with hybrid jowar (r = 0.344) and wheat (r = 0.300) in the basin. Other crops do not denote any significant correlation and hence, these fall in negative correlation. Local jowar has negative correlation with net sown area (r = -0.56). Further the results of correlation has been strengthened by employing multiple regression. It shows high multiple value for cotton and hybrid jowar. Furthermore, earlier approach has been followed up by applying factor analysis techniques to identify unique variance among selected variables of agricultural landuse and environmental variables in the basin in order to search for common variances as a set of new variables. The first four factors account for 83.3 per cent of explained variances which cover cotton factor, density factor, pulses and oilseeds factor, and, wheat and tur factor. First two factors, *namely,* cotton and density show up 43.10 and 20.60 per cent variance and these two factors are employed for obtaining scores. The scores on factor I and II were the base for demarcating agricultural regions into three categories as agricultural backward, developing and developed regions in Yashoda basin. It unfolds more or less dominance of cotton and density factor within the agricultural regions. The study of agricultural regions support to soil types and attitude of farmers i.e. selection of crops, methods for sowing and use of chemical fertilizers.

9

A Study of the Sample Villages

INTRODUCTION

In the preceding Chapters 4 to 8, the efforts were made to assess the spatial analysis of agricultural landuse pattern with an analytical approach applying quantitative measures and 'finally identified the agricultural regions in the Yashoda basin (considering 155 villages) as a whole. Indepth study of every village in the basin is practically difficult and time-consuming. The survey of every parcel of individual village is strenuous and time-consuming and not feasible. Hence the investigator has attempted a study of sample villages, representative of region concerned for indepth and intensive study with a view to evaluate the existing landuse patterns, its problems and prospects.

The objectives put forth for this indepth study of sample villages are as under :

(1) To assess and examine the spatial patterns of agricultural landuse for 1988-89 and interpret the patterns in six sample villages.

(2) To evaluate and examine the temporal analysis of agricultural landuse patterns and find out the variables responsible for this; and,

(3) To identify the problems and prospects of agricultural landuse and suggest measures for proper and efficient landuse to sample villages.

Data readily available in the Land Records of concerned villages for every individual crops have been obtained for chosen villages for the year 1988-89 in both kharif and rabi crops. The data of individual crop was processed and converted into percentage to net sown area and have been used for mapping. In order to understand the temporal changes in crop area, the data of every individual crop were collected, coverted into percentage to net sown area and interpreted. The inaccessible data have been generated by designing questionnaire. The soil samples and field survey in person served to uncover the realistic picture of soil types. The suitable cartographic method used for showing soil types, landuse and crop landuse patterns for the sample villages. The village Patwari, Sarpanch and farmers were the prime source for unfolding the facts related to our aspects of study.

SAMPLING

Sample is recognised as one of the method of selecting representative unit of the region. It is suitable for obtaining factual information pertaining to agricultural patterns and problems of landuse. The desired results can be obtained by indepth study of representative. There is an agreeable fact that survey of selected samples is considerable help if area is of homogeneous in physioco-socio-economic conditions. Sampling provides realistic results at low cost and within minimum time. Samplings are of many types, random, stratified, purposive. The selection mostly depends on purpose of study.

Because of the constraints of the analysis of all villages in the basin, the choice has fallen on six villages. These villages were selected by systematic purposive sampling method and studied at micro level. The crop of each field holding has been plotted on maps of the sample villages by spot inspection and inquiry. The sample villages may be considered as representative of the tracts concerned, mirroring up the agricultural scenario in the chosen year. Exhibit 9.1 shows the location of sample villages in the Yashoda basin and Table 9.1 reveals the population area and density of village concerned.

Table 9.1
Sample Villages

Name of the villages	*Area (in hectares)*	*Population (1989)*	*Density of population*	*Village code number*
Amboda	347.30	475	1.36	34
Nagazari	482.53	1758	3.64	97
Kachangaon	1317.00	1265	0.96	148
Kurzadi	260.87	1258	4.82	53
Sarur	334.37	662	1.97	130
Rohankheda	573.26	780	1.36	141

Source : Author

ASPECTS TO BE STUDIED

For meaningful and indepth analysis of agricultural landuse patterns in sample villages, the following aspects have been considered for study :

(1) Introduction

(2) Climate and Soil

(3) Spatial and temporal analysis of general landuse

(4) Spatial and temporal analysis of agricultural landuse

(5) Landuse and population

(6) Agricultural operation and yields

(7) Socio-economic facilities and farmers.

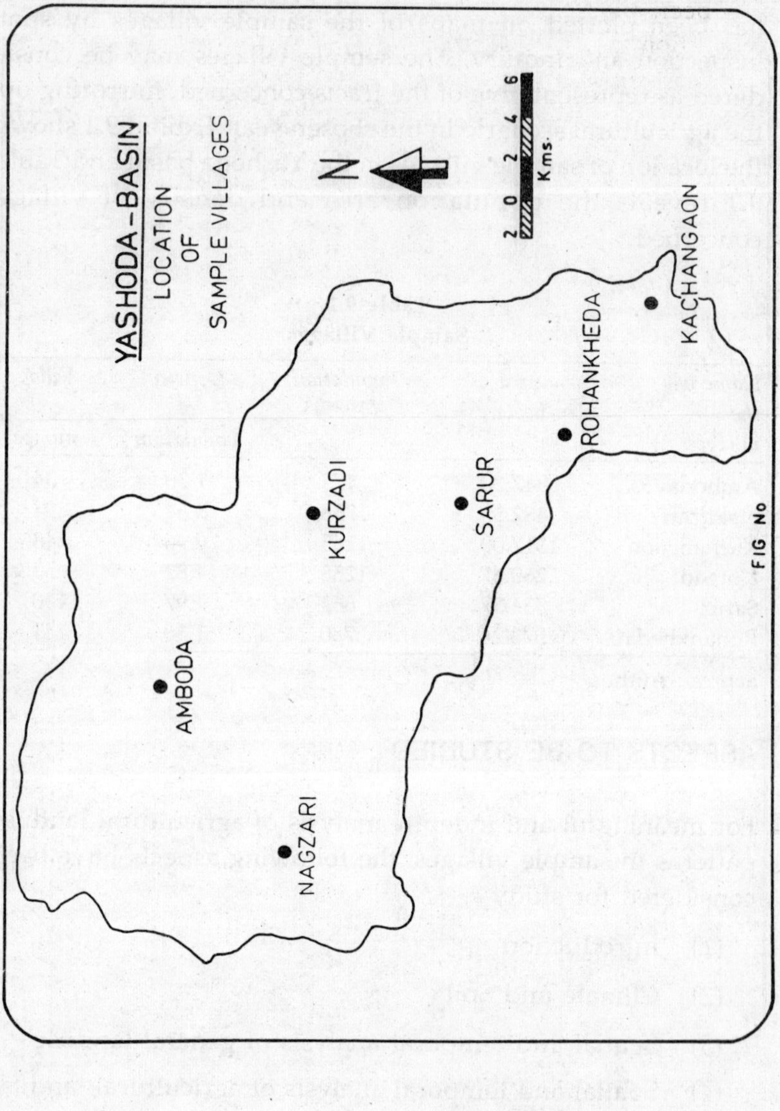
YASHODA-BASIN
LOCATION
OF
SAMPLE VILLAGES
N
2 0 2 4 6
Kms.
KACHANGAON
ROHANKHEDA
KURZADI
SARUR
AMBODA
NAGZARI
FIG No

EXHIBIT — 9.1

QUESTIONNAIRES AND METHOD OF INQUIRY

Two different types of questionnaires have been carefully prepared to obtain the details about agricultural landuse and socio-economic conditions of the farmers. The first set of questionnaire covers the aspects, such as, crops structure, farm implements, sowing methods, selection of seeds, agricultural operation and family background, while, other set of questionnaire was related to village administration to throw light on number of holdings, their sizes, alternative sources of income, farming system, farmers attitude towards the agriculture, if any etc. The interviews were conducted, choosing samples from following land holdings (Table 9.2).

Table 9.2
Land Holding Sizes

Sr. No.	*Size of Land Holdings*
1.	Less than 2 hectares
2.	2 - 4 hectares
3.	4 - 8 hectares
4.	8 -10 hectares
5.	More than 10 hectares

Source : Field work

Besides these, data not readily available and other relevant information on agricultural landuse have been collected from Sarpanch, Patwari. This has been found helpful in throwing light on the aspect under taken for study.

AGRICULTURAL SEASONS

The kharif and rabi are two common seasons of the agricultural year in the study. The kharif is most important season raising nearly 70 per cent crops to total. The crops in kharif are generally sown in June or July and harvested in November and December which include, cotton, hybrid jowar, local jowar, tur, mug, mot, groundnut and kultha. The rabi begins in November and the crops are gram, wheat and linseed.

CLIMATE AND SOILS

The climatic data (rainfall and temperature) are not readily available for sample villages. Hence the investigator relied on the prevailing climatic data, assuming the general climatic conditions to be similar over the tracts, adjoining the meteorological stations, their records have been utilized for the purpose. National Soil Sample Survey and District Soil Survey have prepared soil maps showing broad groups of soil types for Wardha District as a whole. This map does not portray accurate and realistic picture of soil types at the villages level. Hence, investigator, in order to make indepth study of soil types has prepared map of soils for villages concerned based on, soil samples and personal field to field observations during the field work. The local soil classification has been adopted in the present study. This classification covers Kali, Jamin, Bardi and Kharadi Jamin. The soil samples of individual soil types were analysed and soil divisions were demarcated and mapped.

THE SPATIAL AND TEMPORAL STUDY

The spatial as well as temporal analysis of agricultural landuse patterns have been studied for sample villages. The spatial analysis is entirely based on plot to plot field work in person during the year 1988-89 for mapping to depict the general and agricultural landuse patterns. The temporal study is mainly based on Land Revenue Records (1975-76 to 1988-89) and interviews of farmers during the survey.

The longitude and latitude of sample villages have been identified from Indian Topographical Maps. The existing agricultural landuse of individual fields in the choosen villages has been mapped on the scale of 1:2500. The detailed and intensive study of sample villages are as under :

AMBODA

Introduction

Amboda has an areal extent of 347.30 hectares. It is located in the west part in Yashoda basin. The population resides 475 persons in 1989 with density of 1.3 persons per hectare. From the Exhibit 9.2 the surrounding villages can be noticed. It is situated on 70.28' North latitude and 20.46 East longitude. The village is located in north part representative poor development of agricultural activity. The general slope is towards the south which is evident from the flow of water in Nallas (Exhibit 9.2). The gaothan is situated in west in the village area on the left bank of local Nalla. Three Kachha roads appear leading to Dahegaon, Dotra and Nimgaon (Exhibit 9.2).

Climate and Soils

The village experiences monsoon type of climate with low temperature in winter and intensive heat in summer. The rainfall receives from south west monsoon from June to October. The soil of Amboda is coarse shallow with minute variations in texture and colour. There are two types of soils, namely, sandy clay and clay loam. Sandy clay is relatively inferior than clay loam. It lies on extensive area in north, west and east parts covering 73.7 per cent to total village area. Due to low fertility of sandy soil, hardy crops like, mug, local jowar, mot are cultivated. Later soil, namely, clay loam is fertile than earlier, containing 35 per cent clay and spreads it in south parts of villages area. This soil type is suitable for growing cotton and hybrid jowar and consequently, has high yield per hectare. Clay loam cover 25.3 per cent to total area in Amboda (Exhibit 9.2).

Spatial and Temporal Analysis of General Landuse

The general landuse types of Amboda are shown in Table 9.4 and Exhibit 9.3. Out of the total area of 347.30 hectares, 304.81 hectares land is under cultivation accounting for 87.70

per cent to total geographical area. This cultivated land spreads over the area. It restricts area in extreme north and east due to local high relief where land not available for cultivation is predominant. (Exhibit 9.3). Other types of landuse in the village are found isolated in many patches. Among them, large patch in areal extent is observed in north of Amboda gaothan under land not available for cultivation. The total area not available for cultivation is relatively higher than that of other landuse types excluding cultivated land. It accounts for 6.62 per cent to total geographical area followed by cultivable waste (3.65 per cent and fallow land (2.59 per cent). The follow land appears in north part in narrow strip (Exhibit 9.3). It belongs to permanent fallow as this area is dominant of high relief and hence it is not brought under cultivation while cultivable waste was found in two patches. One is the west and other is seen on the right bank of village Nalla near gaothan. This land can be brought under cultivation by levelling. On the whole, village has enough land under plough. The double crops are insignificant accounting for 2 per cent land to total geographical area. This land owned by rich farmers where vegetables and rice are raised.

Table 9.3
Soil Analysis

Soil Texture Classes	*Soil Types*	
	Sandy clay (%)	*Clay loam (%)*
Coarse sand	15.00	11.60
Fine sand	30.76	25.26
Silt	20.00	23.75
Clay	28.00	35.00
Nitrogen	0.38	0.31
Phosphate	—	2.08
Potash	58.56	8.25
pH Value	7.7	7.8

Note : Above analysis is carried out in Soil Testing Laboratory, Agricultural Department, Pune-5.

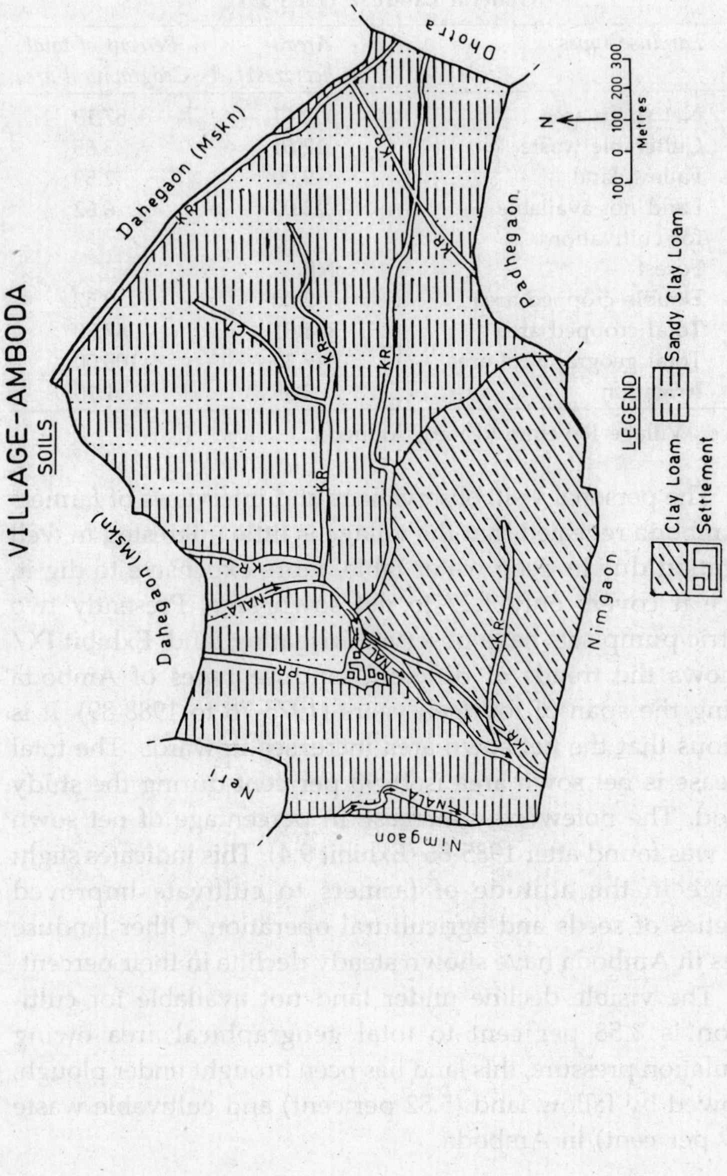
VILLAGE-AMBODA
SOILS
Dahegaon (Mskn)
Dhotra
Padhegaon
Nimgaon
Neri
KR
NALA
PR
N
100 0 100 200 300
Metres
LEGEND
Clay Loam
Sandy Clay Loam
Settlement

EXHIBIT — 9.2

Table 9.4
General Landuse (1988-89)

Sr. No.	*Landuse types*	*Area (in hectares)*	*Percent of total Geographical area*
1.	Net sown area	302.61	87.13
2.	Cultivable waste	12.61	3.65
3.	Fallow land	9.00	2.59
4.	Land not available for cultivation	23.00	6.62
5.	Forest	Nil	Nil
6.	Double cropped area	2.00	0.57
7.	Total cropped area	304.61	87.70
8.	Total geographical area	347.30	100.00
9.	Irrigation	3.85	1.10

Source : Village Revenue Record, Amboda.

The personal visit, observations and interviews of farmers of Amboda revealed that the village is little interested in well irrigation due to deep water table, more expensive to dig it, hence it covers only 1.27 to net sown area. Presently two electric pump sets have observed irrigating land. Exhibit IX/5 shows the trends of different landuse types of Amboda during the span of fourteen years (1975-76 to 1988-89). It is obvious that the net sown area increases upwards. The total increase is net sown area is 16.98 per cent during the study period. The noteworthy increase in percentage of net sown area was found after 1985-86 (Exhibit 9.4). This indicates slight change in the attitude of farmers to cultivate improved varieties of seeds and agricultural operation. Other landuse types in Amboda have shown steady decline in their percentage. The visible decline under land not available for cultivation is 8.58 per cent to total geographical area owing population pressure, this land has been brought under plough, followed by fallow land (5.52 per cent) and cultivable waste (2.89 per cent) in Amboda.

Table 9.5

Temporal Variations in Landuse (1975-76 to 1988-89)

Years	*General Landuse (Per cent of total geographical area)*					*Agricultural Landuse (% of net sown area)*						
	NSA	CW	FL	ANC	Forest	Cotton	Local Jowar	Hybrid Jowar	Other cereals	Total Pulses	Total oil	Other miscellaneous seeds
1975-76	70.15	6.54	8.11	15.20	NIL	30.70	40.20	0.00	4.19	21.00	2.11	1.80
1976-77	68.50	6.20	10.30	15.00	NIL	28.10	35.10	0.00	7.52	19.30	4.87	5.11
1977-78	68.35	7.75	8.60	14.30	NIL	27.30	35.50	5.30	7.30	17.11	4.42	2.98
1978-79	68.33	8.55	8.62	14.50	NIL	31.33	33.70	7.10	6.77	15.72	3.55	1.83
1979-80	66.67	8.78	10.65	14.00	NIL	32.55	31.50	7.80	8.32	13.11	5.19	1.53
1980-81	67.50	8.90	9.92	13.68	NIL	35.19	30.15	8.15	8.85	13.30	4.36	Nil
1981-82	69.38	9.20	8.25	13.17	NIL	37.39	25.25	8.25	6.53	13.90	4.71	3.97
1982-83	70.52	9.70	7.28	12.50	NIL	38.25	21.30	10.42	7.10	13.25	5.37	4.31
1983-84	72.75	8.30	7.20	11.95	NIL	40.11	19.25	12.15	6.73	14.00	4.20	3.46
1984-85	73.15	7.50	9.05	10.30	NIL	42.52	12.17	15.37	5.30	14.10	8.11	2.43
1985-86	75.50	6.75	6.65	10.50	NIL	44.30	11.37	17.30	4.45	14.20	5.00	3.00
1986-87	79.51	5.78	6.12	8.59	NIL	44.68	8.25	21.22	4.10	13.08	4.25	4.42
1987-88	84.52	4.37	3.79	7.32	NIL	46.32	5.10	25.85	3.06	13.80	4.10	1.77
1988-89	87.13	3.65	2.59	6.62	NIL	49.56	3.58	29.74	0.87	12.98	2.19	0.91

Source : Village Revenue Records, Amboda

Note : NSA - Net Sown Area, ANC - Area Not Available for Cultivation, FL - Follow Land, CW - Cultivable Land.

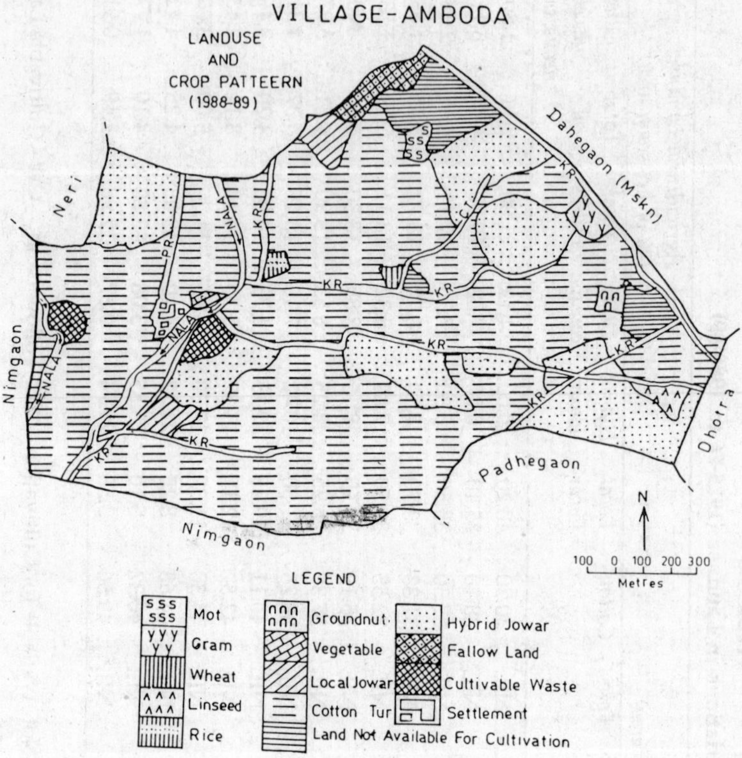
VILLAGE-AMBODA
LANDUSE
AND
CROP PATTEERN
(1988-89)
Neri
Nimgaon
Nimgaon
Dahegaon (Mskn)
Padhegaon
Dhotra
NALA
KR
PR
N
100 0 100 200 300
Metres
LEGEND
Mot
Gram
Wheat
Linseed
Rice
Groundnut
Vegetable
Local Jowar
Cotton Tur
Land Not Available For Cultivation
Hybrid Jowar
Fallow Land
Cultivable Waste
Settlement

EXHIBIT — 9.3

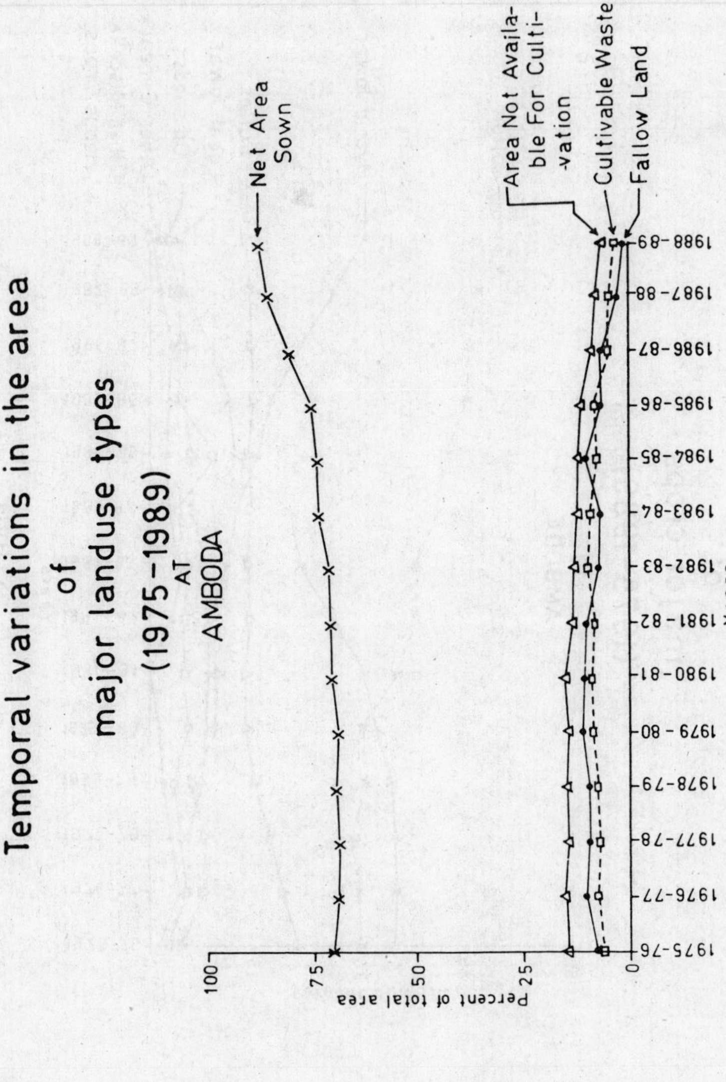
Temporal variations in the area
of
major landuse types
(1975–1989)
AT
AMBODA
Net Area Sown
Area Not Available For Cultivation
Cultivable Waste
Fallow Land
Percent of total area
100
75
50
25
0
Years
1975-76
1976-77
1977-78
1978-79
1979-80
1980-81
1981-82
1982-83
1983-84
1984-85
1985-86
1986-87
1987-88
1988-89

EXHIBIT — 9.4

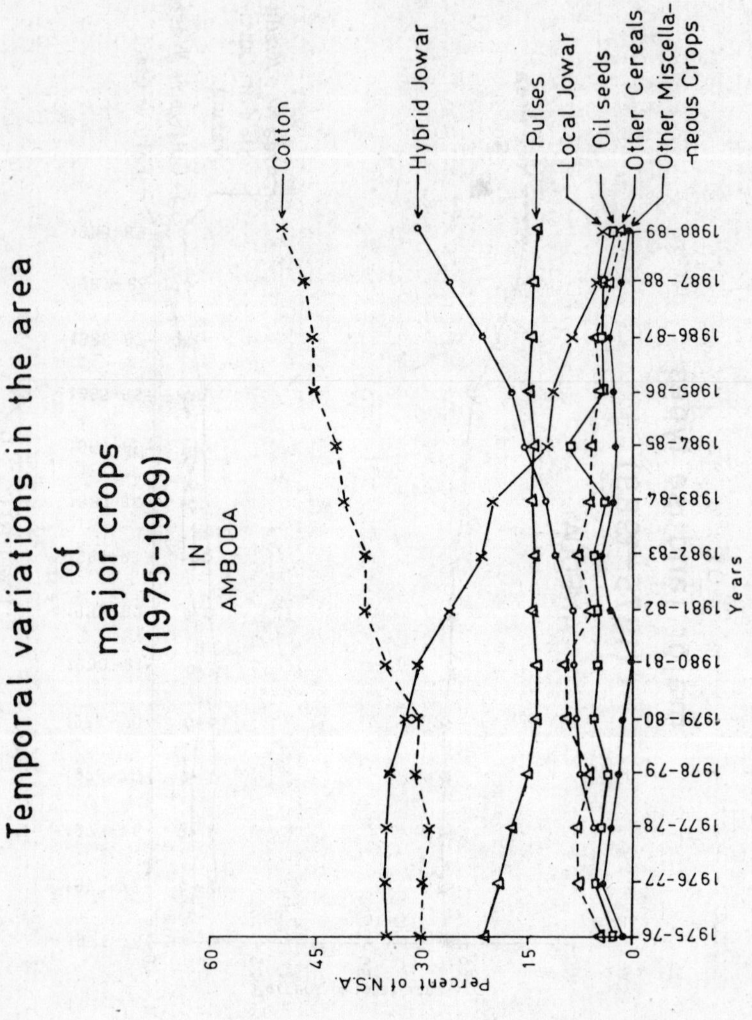
Temporal variations in the area
of
major crops
(1975-1989)
IN
AMBODA
Cotton
Hybrid Jowar
Pulses
Local Jowar
Oil seeds
Other Cereals
Other Miscella-neous Crops
Percent of N.S.A.
60
45
30
15
0
1975-76
1976-77
1977-78
1978-79
1979-80
1980-81
1981-82
1982-83
1983-84
1984-85
1985-86
1986-87
1987-88
1988-89
Years

EXHIBIT — 9.5

Spatial and Temporal Analysis of Agricultural Landuse

Amboda has 304.61 hectares under cultivation. Out of this, 0.57 percentage double cropping. Irrigation is found on 1.10 per cent land to net sown area. The crops are sown in two seasons; *namely*, kharif and rabi. Cotton, hybrid jowar, local jowar, mug, mot, sesamum. Among these, cotton and hybrid jowar are raised on large areal extent accounting for 49.56 per cent and 29.74 per cent to net sown area respectively. The cotton is on extensive hectarage, well-spread in the village than hybrid jowar (Exhibit 9.3). Hybrid jowar is found in many patches, sown on both sandy clay and clay loam soils. The major patch of hybrid jowar, in areal extent, appears along the road which leads towards the east. Among other crops, tur stands third denoting 11.58 per cent land to net sown area in cotton devoting field sown area. Local jowar, mot, mug are mainly raised on sandy loam soil, resulting low yield per hectare. Wheat accounts for 0.82 per cent to net sown area raised in rabi season. Other crops raised in rabi period are, namely, gram and linseed, covers 0.66 and 0.14 per cent land to net sown area in Amboda.

Table 9.5 shows the temporal variations in agricultural landuse pattern in Amboda from 1975-76 to 1988-89. Two crops, *namely*, cotton and hybrid jowar have steady upward trend during study period. It is clear from Table 9.5 that the rate of increase in per cent of cotton is comparatively high than that of hybrid jowar. This indicates the dominance of cotton after 1980-81. The overall increase in per cent under cotton is 18.86 per cent to net sown area. Cotton had grown on 30.70 per cent to net sown are in 1975-76 while it was 49.56 per cent to net sown area (1988-89). This significant increase in per cent under this crop is attributed to cultivate new improved varieties of cotton for sowing and therefore cotton has remained first in rank in Amboda. The similar trend is also observed in case of hybrid jowar for the span of fourteen (Exhibit 9.5) years. It increased by 14.35 per cent and has wide distribution in the village. Among other crops, local jowar presents remarkable decline trend in Amboda than that of other crops *namely*, pulses, other cereals and oilseeds. The total decline of local jowar is 10.46 per cent due to low yield per hectare. It had registered 40.20 per cent to net sown area in 1975-76 which is declined to 29.74 per cent to net sown area in 1988-89. The area under pulses is more or less constant in the village. This trend is continued for oilseeds and other pulses. Pulses declined by 16.02 per cent to net sown area followed by other cereals (3.32 per cent), other miscellaneous crops (0.89 per cent) and oilseeds (0.08 per cent). The cropping pattern in the span of fourteen years dominated by cotton and hybrid jowar.

Table 9.6

Agricultural Landuse (1988-89)

Sr. No.	*Name of crops*	*Percent to net sown area*	*Sr. No.*	*Name of crops*	*Percent to net sown area*
1.	Cotton	49.56	7.	Chilli	0.38
2.	Hybrid jowar	29.74	8.	Vegetables	0.44
3.	Local Jowar	3.58	9.	Wheat	0.82
4.	Tur	11.56	10.	Gram	0.66
5.	Rice	0.05	11.	Groundnut	2.19
6.	Mot	0.76	12.	Linseed	0.14

Source : Village Revenue Records, Amboda.

Landuse and Population

Table 9.7, shows the landuse and population in Amboda. It is clear from this table that population has relatively more pressure on land available for cultivation and therefore land available for cultivation shows 0.63 hectare per head of population. Furthermore, it has been observed that area under kharif crops has more land per head of population than that of rabi crops in the village. It is 0.61 hectare per head of population in area under kharif while 0.02 hectare per hand of population is in rabi.

Table 9.7
Landuse and Population

(Population 475)

Sr. No.	Landuse categories	Area (hectares)	Land per head of population
1.	Total village area	347.30	0.73
2.	Land available for cultivation	302.61	0.63
3.	Total kharif crops area	291.16	0.61
4.	Total rabi crops area	11.45	0.02
5.	Gross cropped area	304.61	0.64
6.	Double cropped area	2.00	0.05

Source : Author

This fact explains that kharif area has dominance and pressure of population in Amboda. The demographic characteristics are shown in Table 9.8. The population density increases in Amboda through time. The density increased by 0.31 persons per hectare during the period of 1971 to 1989. It was registered 1.05 persons per sq km in 1971 while it is noticed 1.36 persons per sq. km in 1989 due to population growth. The working force has also showed upward trend and increased by 5.55 per cent to total workers. The increase in working force uncovers the increasing pressure of population on land in Amboda and therefore per capita land in the village is very low (0.73). Literacy aspect in Amboda presents more or less a static picture. The cultivators have increased in 40.64 per cent and 76.00 per cent to total workers. Moreover trend of 5.37 per cent from 1971 to 1989. These facts explain the dominance of agriculture activity in the village (Table 9.9).

Table 9.8
Demographic Characteristics

Sr. No		Years 1971	1981	1989
1.	Total Population	365	399	475
2.	Density of Population	1.05	1.14	1.36

3.	Literacy percentage	36.35	56.82	49.47
4.	Per capita land	0.95	0.87	0.73
5.	Percentage of farm workers to total workers	89.02	95.20	96.57

Source : Wardha District Census Handbook for 1971 and 1981, and Village Panchayat Office, Amboda.

Table 9.9

Occupational Structure

Sr. No.	Categories	Population			Percent of total population			Percent of total population		
					Years					
		1971	*1981*	*1989*	*1971*	*1981*	*1989*	*1971*	*1981*	*1989*
1.	Cultivators	63	100	133	17.26	25.06	28.00	40.64	80.00	76.00
2.	Agricultural Labourers	75	19	36	20.54	4.76	7.57	48.38	15.20	20.57
3.	Marginal workers	-	-	-	-	-	-	-	-	-
4.	Workers other than farm workers	17	6	6	4.65	1.50	1.26	10.96	4.8	3.42
5.	Total workers	155	125	175	42.46	31.22	3684	100.0	100.0	100.0
6.	Non-workers	210	274	300	57.53	68.67	63.15	—	—	—
7.	Total Population	365	399	475	100.00	100.0	100.0	—	—	—

Source : District Census Handbook for 1971, 1981 and Village Panchayat Office, Amboda.

Agricultural Operation and Yields

Agricultural operation and yields of major crops are shown in Table 9.10. It was noticed that cotton is dominant crop, cultivated on extensive area (49.58 per cent to net sown area) on clay loam and sandy clay soils. It is sown in June or July depending on commencing monsoon. It is mixed with tur and sown. The average yield of cotton in Amboda in 150 to 200 kilograms. Hybrid jowar is rainfed crop and is sown by drilling method yielding 100 to 150 kilogram. Local jowar, tur, and groundnut are sown on all soil types and their yields mostly depends on coverage of these crops in field. It is found that local jowar has still significance hectarage in Amboda.

These three crops are rainfed and sown by drilling method (Table 9.10).

Socio-Economic Facilities and Farmers

The majority of farmers are economically poor and therefore farmers could not significantly improve their agricultural operation. Small farmers do not possess tractors, improved seeds or iron plough. They are even unable to invest money on implements and has hindrance in improving agriculture. The needed capital is not readily available to farmers. Because of lack of capital the farmers cannot purchase improved varieties of seeds. Other aspect, size of holding affects agriculture. It was noticed that majority of farmers have less than two hectares land disintegrated in various pieces and therefore efficient landuse is lacking. Small size land holders are not possessed of improved implements. It is found that small farmers are borrowing or hired implements from neighbours. Moreover, absence of post office, high school, somewhat have adversed affect on existing agricultural pattern in Amboda.

Resume

Agriculture in Amboda is backward due to many reasons, firstly, more than 70 per cent land is sandy loam where yield per hectare is low secondly, economic condition of farmers are poor, and therefore farmers cannot purchase improved implements, seeds, and pesticides. Cotton and hybrid jowar are major crops covering 49.56 per cent and 29.74 per cent to net sown area respectively. Other crops are of little significance as they altogether cover nearly less than 30 per cent land to net sown area in the village. Tur is sown, mixed with cotton. It is noticed that agricultural labours and cultivators are increased from 1981 to 1989. Irrigation is practiced on 1.10 per cent area to net sown area. Lack of communication, post office, high school and shortage of water indicates the backwardness of the village.

Table 9.10
Agricultural Operation and Yields

Landuse Operations	*Crops*				
	Cotton	*Hybrid Jowar*	*Tur*	*Local Jowar*	*Groundnut*
Botanical Name	Gossypium Spp.	Sorghum bicolor	Cajanus indicus	Sorghum vulgare	Arachia hypogaea Linn.
Soil	Clay loam/ sandy clay	Clay loam/ sandy clay	Clay loam/ sandy clay	Clay loam/ sandy clay	Sandy clay
Sowing time	June/July	June	June/July	June	June/July
Spacing	3 to 4 1/2'	7 to 10"	3 to 4 1/2'	7 to 9'	5 to 7'
Seed rate per acre in Kg	1 Kg	1 Kg	2 Kg	2 Kg	2 1/2 Kg
Method of sowing	Dibbling	Drilling	Dibbling	Drilling	Drilling
Manures of Ferti-lizers (NPK) in Kg. per acre	35-7-0	10-2-0	Nil	Nil	Nil
Season	Rainfed (Kharif)	Rainfed (Kharif)	Rainfed (Kharif)	Rainfed (Kharif)	Rainfed (Kharif)
Harvesting time	October-January	October	January	October	October
Yield per acre in Kg.	150-200 Kg	100-150 Kg.	100- 125 Kg.	100 Kg.	125 kg.

Source : Field work

2. NAGAZARI

Introduction

Nagazari is located in west part in Yashoda basin (Exhibit 9.1). It is away from 10 Kms from Dewali and 28 Kms from Wardha. According to data of Gram Panchayat, population of village is 1738 persons and density of 3.60 persons per hectare. Nagazari occupies 482.53 hectares. The gaothan is in centre (Exhibit 9.6). The village is confined by Muradgaon in east, Belgaon in west, Khadaki and Palasgaon in east. The general slope is towards east. It is evident from water flow of Nalla (Exhibit 9.6), the majority of Kunbi appears involving in agricultural activity, particularly as cultivators, while minority people, namely, Mahar, Chambar, and others are working as agricultural workers.

Climate and Soil

Intensive heat in summer and cold winter is observed in Nagazari. June to October receiving rainfall from southwest mansoon from June to October prove useful for kharif crops. The personal observations during field work and soil analysis reveals three soils types, *namely*, sitly loam, clay loam and clay. These soil types are locally known as Panthari, Kali and madhyam soils. The analysis of soil types of Nagazari is given in Table 9.11. Among these soils types, silty loam covers 45 per cent to total village area. This soil is relatively fertile than that of other two types containing highest percentage of silt (52.50). Clay loam is found on extensive area in west part of the village on 33 per cent to total area. This soil is black in colour and productive. It contains 27.50 per cent silt and 35.50 per cent clay (Table 9.11). Third type of soil, *namely*, clay is found in the south and southeast to gaothan (Exhibit 9.6). It covers 22 per cent to total geographical area.

Table 9.11
Soil Analysis

Soil texture classes	*Soil Types*		
	Silty loam (per cent)	*Clay loam (per cent)*	*Clay (per cent)*
Coarse sand	13.70	8.10	9.80
Fine sand	13.23	24.26	24.69
Silt	52.50	27.50	10.00
Clay	18.00	35.50	50.50
Nitrogen	0.29	0.55	50.50
Phasphate	1.72	1.16	1.60
Potash	15.00	23.25	12.25
pH value	8.1	8.0	8.1

Note : The above soil analysis was carried out in soil Testing Laboratory, Agriculture Department, Pune-5.

Spatial and Temporal Analysis of General Landuse

The landuse types under different categories are shown in Table 9.12. Nagazari has 328.40 hectares (68.05 per cent to total geographical area) under cultivation. The concentration of cultivated land is mainly related to the fertility of soils and therefore major concentration of cultivated area has been on fertile track in the easternpart in Nagazari. The concentration of cultivated land slightly declined towards west as fertility of soil decreases. The land not available for cultivation is 16.60 per cent to total geographical area lying in western boundary of the village which cannot be brought under cultivation due to existence of local high relief and stony surface and restricts to extent cropland. This is part of barren and open with disperse shunted trees. The fallow land covers 9.84 per cent to total geographical area in Nagazari in two patches disintegrated from each. The major extensive patch of this land is found in north part (Exhibit 9.7) followed by cultivable waste, accounting for 5.49 per cent to total geographical area. It is observed that the fallowed land has deliberately kept under this category in order to region fertility in northeast part. The double crop area in Nagazari is insignificant on 2.35 per cent to total geographical area and has found in the eastern part on the roadside. (Exhibit 9.7) whereas, 1.63 per cent to total geographical area is under irrigation.

Table 9.12

General Landuse Pattern (1986-89)

Sr. No.	*Landuse types*	*Area (hectares)*	*Percent of total geographical area*
1.	Net sown area	328.40	66.05
2.	Cultivable waste	26.50	5.49
3.	Fallow Land	49.50	9.84
4.	Land not available for cultivation	80.13	16.60
5.	Forest	Nil	Nil
6.	Double cropped area	11.37	2.35
7.	Total cropped area	33.77	70.40
8.	Irrigated area	7.90	1.63
9.	Total geographical area	462.53	—

Source : Village Revenue Records, Nagazari.

Exhibit 9.7 and Table 9.13 present temporal variations in general landuse pattern in Nagazari from 1975-76 to 1986-89. The area in per cent under net sown area has been slightly increased by 6.40 per cent. This increase in net sown area is comparatively less than of other sample villages. The net sown area had 61.65 per cent to total geographical area in 1975-76 which was increased to 68.05 per cent in 1988-89. The less increase under net sown area due to extensive land is under the category of land not available for cultivation. Therefore, this land was not brought largely under cultivation. The overall decline in land not available for cultivation was 4.98 per cent during fourteen years. Cultivable waste was not brought under plough therefore this land shows little decline (2.16 per cent to total geographical area). The fallow land has found increase by 0.74 per cent to total geographical area. It had 9.10 per cent land under fallow to total geographical area

Table 9.13
Temporal Variations in Landuse (1975-76 to 1988-89)

Years	*General Landuse (Per cent of total geographical area)*					*Agricultural Landuse (% of net sown area)*						
	NSA	*CW*	*FL*	*ANC*	*Forest*	*Cotton*	*Local Jowar*	*Hybrid Jowar*	*Other cereals*	*Total Pulses*	*Total oil*	*Other miscella-neous seeds*
1975-76	61.65	7.67	9.10	21.58	NIL	28.12	32.25	NIL	23.27	21.13	16.36	NIL
1976-77	61.20	12.13	6.17	20.50	NIL	26.55	32.10	NIL	11.10	20.20	9.95	NIL
1977-78	60.52	10.10	9.25	20.13	NIL	26.15	30.52	NIL	10.34	19.17	6.71	7.11
1978-79	62.11	10.59	7.30	20.00	NIL	26.30	27.15	NIL	12.38	17.25	9.94	6.98
1979-80	63.70	9.27	9.84	17.19	NIL	28.55	22.25	5.35	12.52	15.42	10.00	5.91
1980-81	63.10	10.32	8.33	18.25	NIL	30.48	20.38	7.10	11.35	14.12	10.42	6.15
1981-82	64.78	9.15	8.78	17.29	NIL	32.11	18.55	11.22	10.42	13.27	8.78	5.65
1982-83	64.23	9.00	9.42	17:35	NIL	34.22	12.62	15.29	14.31	13.85	9.71	NIL
1983-84	65.19	12.50	16.61	15.12	NIL	37.18	10.11	18.26	10.53	13.10	5.71	5.11
1984-85	65.42	10.15	6.08	18.35	NIL	38.00	10.77	23.77	8.00	12.90	3.87	3.20
1985-86	65.32	6.46	7.69	10.53	NIL	39.11	9.27	25.30	7.75	12.87	2.58	3.12
1986-87	65.63	10.78	11.40	12.19	NIL	43.27	7.31	27.11	5.98	12.92	2.41	1.00
1987-88	66.18	9.10	9.52	15.20	NIL	45.19	7.14	28.25	4.10	11.20	2.80	1.32
1988-89	68.05	5.49	9.84	16.60	NIL	46.49	5.99	30.90	3.50	11.16	0.90	1.00

Source : Village Revenue Records, Nagazari.

Note : NSA - Net Sown Area, ANC - Area Not Available for Cultivation, FL - Follow Land, CW - Cultivable Land.

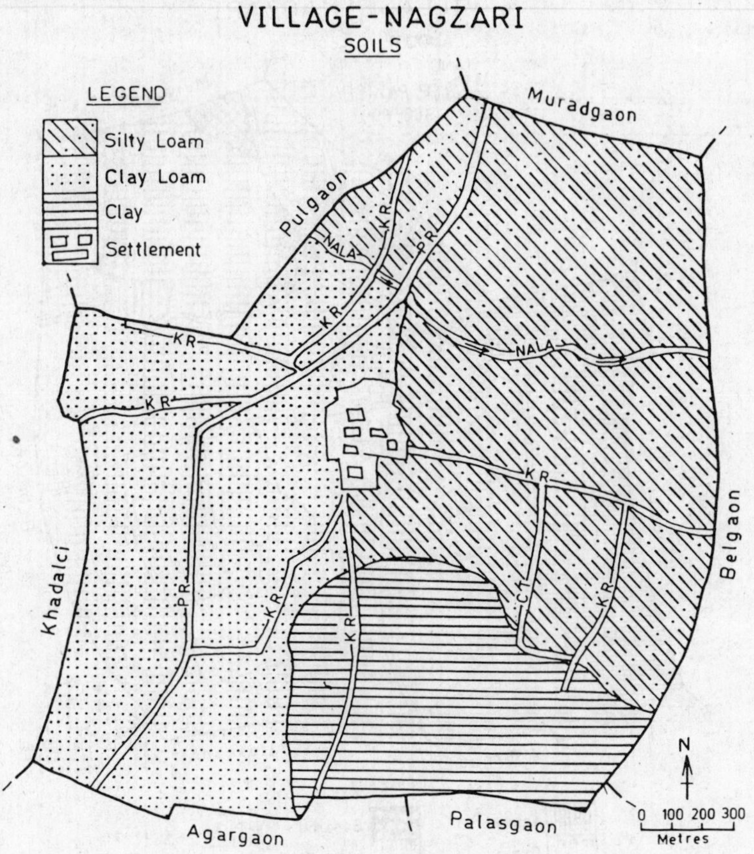
VILLAGE-NAGZARI
SOILS
LEGEND
Silty Loam
Clay Loam
Clay
Settlement
Muradgaon
Pulgaon
NALA
KR
PR
KR
KR
NALA
KR
KR
Belgaon
Khadalci
PR
KR
KR
CT
KR
N
0 100 200 300
Metres
Agargaon
Palasgaon

EXHIBIT — 9.6

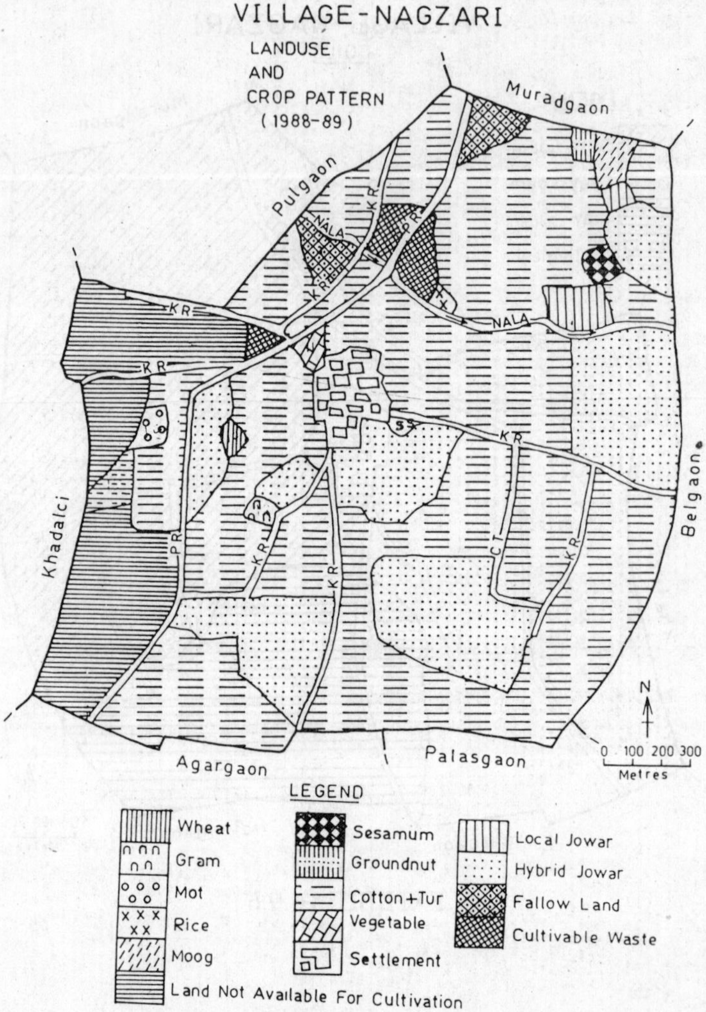
VILLAGE-NAGZARI
LANDUSE
AND
CROP PATTERN
(1988-89)
Muradgaon
Pulgaon
Khadalci
Belgaon
Agargaon
Palasgaon
NALA
KR
PR
CT
N
0 100 200 300
Metres
LEGEND
Wheat
Gram
Mot
Rice
Moog
Land Not Available For Cultivation
Sesamum
Groundnut
Cotton+Tur
Vegetable
Settlement
Local Jowar
Hybrid Jowar
Fallow Land
Cultivable Waste

EXHIBIT — 9.7

Temporal variations in the area of major landuse types (1975-1989) AT NAGAZARI

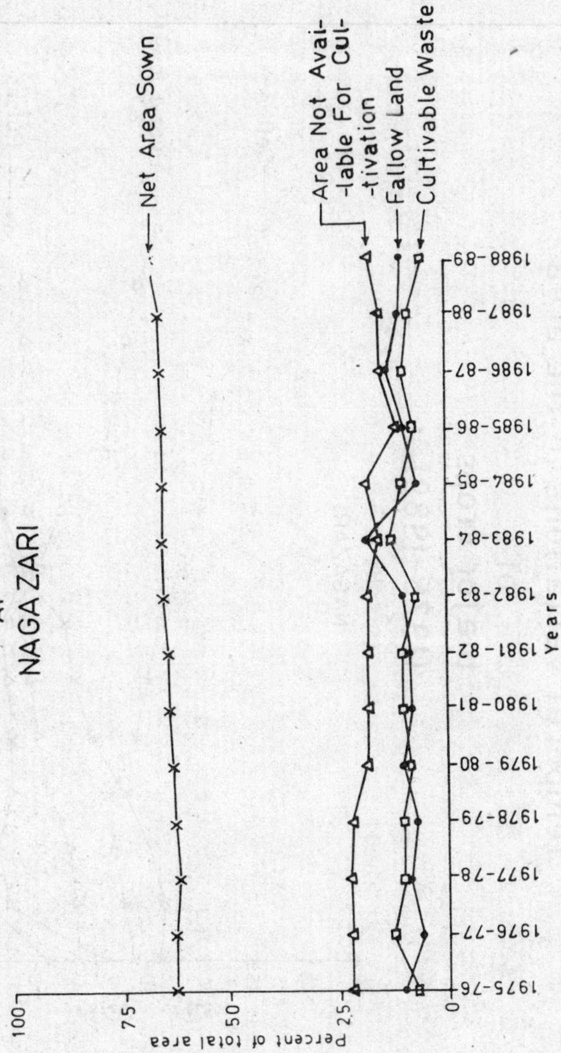

EXHIBIT — 9.8

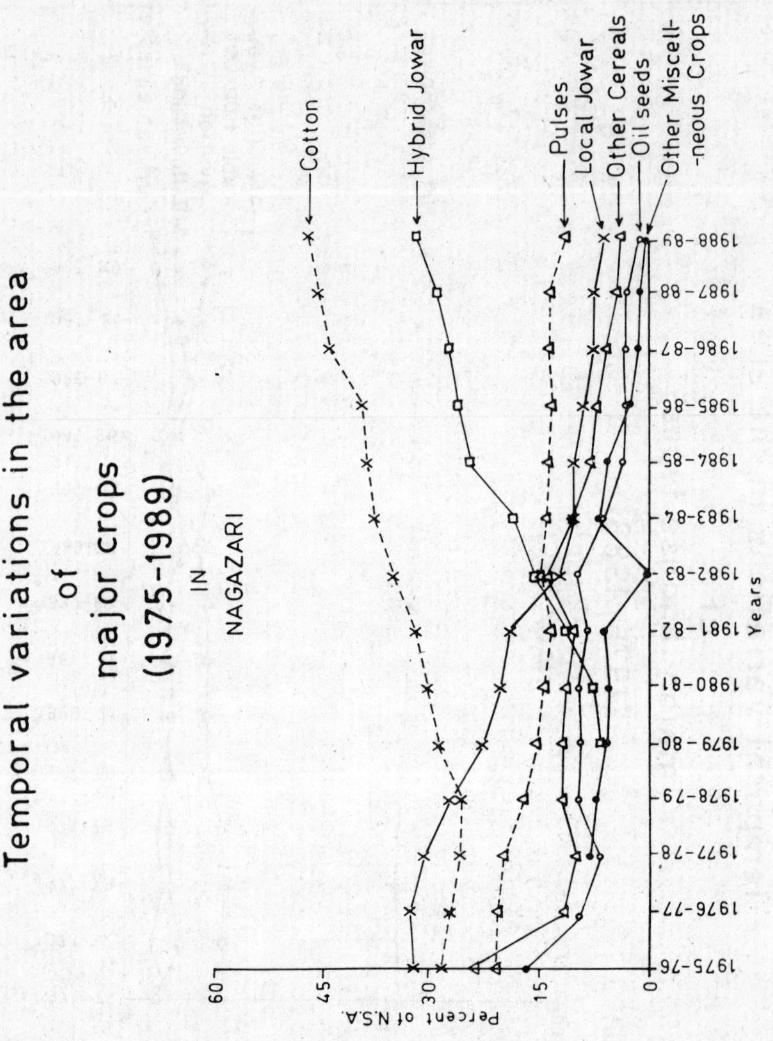
Temporal variations in the area
of
major crops
(1975-1989)
IN
NAGAZARI
Cotton
Hybrid Jowar
Pulses
Local Jowar
Other Cereals
Oil seeds
Other Miscell-
-neous Crops
Percent of N.S.A.
60
45
30
15
0
1975-76
1976-77
1977-78
1978-79
1979-80
1980-81
1981-82
1982-83
1983-84
1984-85
1985-86
1986-87
1987-88
1988-89
Years

EXHIBIT — 9.9

which was attained 9.64 per cent in 1988-89. Forest cover is almost absence in Nagazari.

Spatial and Temoral Analysis of Agricultural Landuse Pattern

68.05 per cent to total geographical area is under cultivation and 80.69 per cent farm workers in Nagazari. This net sown area is comparatively less than that of sample villages. Agriculture in Nagazari is still traditional approach to agricultural practice. Soil type is largely influenced on existing agricultural landuse pattern. Generally, fertile land is devoted to cotton and hybrid jowar (Exhibit 9.7). Both cotton and hybrid jowar are dominant crops covering 46.49 per cent and 30.90 per cent to net sown area respectively. The major concentration of cotton is observed in north, central and south parts on clay and silty loam soils and consequently produce high yield per hectare, while the concentration of hybrid jowar is seen in the south and east in three patches on all soil types. Tur ranks third in cropping pattern, accounting for 8.74 per cent to net sown area followed by local jowar (5.99 per cent to net sown area). Tur is mixed crop sown with cotton in two rows after every ten rows of cotton. The inferior soil is occupied by local jowar in Nagazari in the north part ranking fourth in cropping. It accounts for 5.94 per cent to net sown area. Among rabi crops, wheat occupies larger areal extent of 3.35 per cent to net sown area in the west of gaothan on clay loam soils, while gram is sown on small patch overing 0.03 per cent to net sown area. Individually crop, *namely*, sesamum, mug, rice, groundnut, chilly and vegetables do not cover more than 1.00 per cent to net sown area and therefore these crops are sown on inferior soils in the village.

Table 9.13 and Exhibit 9.9 display temporal variations in agricultural landuse pattern in Nagazari for the span of fourteen years (1975-76 to 1988-89). The village has dominance of agricultural activity involving 80.69 per cent working force to total workers in agriculture. Exhibit 9.9 shows similar upward trend, like other sample villages for cotton and hybrid jowar. This indicates slight positive change in improving agriculture. But overall progress of agricultural landuse is more or less constant. It is clear from Table 9.13, that cotton and hybrid jowar have increased by 18.37 per cent and 25.55 per cent to net sown area respectively. The rate of increase of hybrid jowar is relatively high than that of cotton (Table 9.13). This fact explains positive attitude of farmers to prefer high yielding varieties of hybrid jowar in the place of local jowar and therefore areal extent has remarkably declined in Nagazari. It declined by 26.26 per cent during study period. It had 32.25 per cent to net sown area under local jowar in 1925-76 which is registered 5.99 per cent to net sown area in 1988-89. The Exhibit 9.9 shows significant decline under local jowar. The area under crops, namely, oilseeds, other cereals and pulses has shown downward trend with an individual decline of less than 22 per cent to net sown area. The highest decline is observed in other cereals which is 20.63 per cent to net sown area followed by oilseeds (15.30 per cent) pulses (9.37 per cent) and other miscellaneous crops (6.11 per cent) in Nagazari.

Table 9.13/A
Agricultural Landuse Pattern (1988-89)

Sr. No.	*Name of crops*	*Percent to net sown area*	*Sr. No.*	*Name of crops*	*Percent to net sown area*
1.	Cotton	46.49	8.	Mot	1.61
2.	Hybrid jowar	30.90	9.	Chilli	0.57
3.	Local jowar	5.99	10.	Vegetables	0.43
4.	Tur	8.74	11.	Wheat	3.35
5.	Sesamum	0.78	12.	Gram	0.03
6.	Mug	0.78	13.	Groundnut	0.12
7.	Rice	0.15			

Source : Village Revenue Records, Nagazari

Landuse and Population

Table 9.15 summerises the per capita share of land to various categories of landuse in Nagazari. The per capita land is 0.27 hectare. The per capita land available for cultivation is 0.18 hectare. Further it is reduced to 0.18 hectare per capita and in area under kharif crops and 0.06 in rabi crops area. This fact reflects dominance of kharif hectarage in the village. The population has steadily increased from 1971 (Table 9.5). It was 1771 persons in 1911 while 1738 persons in 1989 with density of 3.54 and 3.60 persons per hectare respectively. Literacy has not found increase significantly. 82.69 per cent working force is engaged in agriculture. This suggests dominance of agriculture. The occupational structure of Nagazari is presented in Table 9.17 showing population engaged in various occupations proportion to total population and total workers. Both cultivators and agricultural labourers are increased in number between 1971 and 1989. Cultivators have increased from 166 to 261 while agricultural labourers have registered an increase from 485 to 580 in Nagazari. Total increase of cultivators is 4.19 per cent to total workers. The increase in number under agricultural labourers indicates involvement in single agricultural activity.

Agricultural Operation and Yields

The agricultural operation and yields of major crops are shown in Table 9.18. These major crops are sown on clay, clay loam and silty loam soils. Cotton, hybrid jowar are cultivated in kharif season while wheat is from in rabi season. Clay and clay loam is preferred for cotton cultivation. Drilling method is common for sowing. The maturity time to crop vary. Cotton takes four to five months while tur matures within four months. Cotton and hybrid jowar require chemical fertilizers and sprayings of pesticides. Wheat is major crop in rabi season and is sown by drilling method. The per hectare yield of cotton is higher than of other crops (Table 9.18). The yield of cotton mainly depends on soil types, application of fertilizers and timely sprayings of pesticides. Local jowar and wheat do not require fertilizers.

Table 9.15
Landuse and Population

(Population 1738)

Sr. No.	Landuse categories	Area (hactare)	Land per head of population
1.	Total village area	482.53	0.27
2.	Land available for cultivation	328.40	0.18
3.	Total Kharif crops area	316.67	0.18
4.	Total rabi crops area	11.53	0.06
5.	Cross cropped area	339.77	0.19
6.	Double cropped area	11.37	0.06

Source : Author

Table 9.16
Demographic Characteristics

Sr. No.	Categories	Years 1971	1981	1989
1.	Total Population	1711	1667	1738
2.	Density of Population	3.54	3.45	3.6
3.	Literacy percentage	41.14	52.60	55.06
4.	Per capita land	0.28	0.29	0.27
5.	Percentage of farm	83.95	91.35	82.69

Source : Wardha District Census Handbook for 1971 and 1961, and Village Panchayat Office, Nagazari.

Socio-Economic Facilities and Farmers

It is noticed that nine farmers are having more than four hectare land, while rest owned less than two hectares. The small size land holders cannot cultivate land properly and efficiently as they are economically poor. They cannot afford improved seeds and fertilizers. The size of land is very small attributed to constant family divisions and subdivisions. Moreover, these farmers are, by and large, conservative in their attitude. Further it was noticed that two farmers have more than 10 hectares land. These farmers cultivate improved seeds and fertilizers largely and get high return. Credit facilities are not locally available to farmers. One has to travel to Dewali to avail himself of these facility. Local money lenders at Nagazari provide credit to farmers.

Table 9.17
Occupational Structure

Sr.	Categories	Population			Percent of total population			Percent of total population		
					Years					
		1971	1981	1989	1971	1981	1989	1971	1981	1989
1.	Cultivators	166	243	261	9.70	14.57	15.01	21.47	25.49	25.66
2.	Agricultural labourers	485	560	580	28.34	33.59	33.37	62.74	58.76	57.03
3.	Marginal workers	101	74	72	5.90	4.43	4.14	13.06	7.76	7.07
4.	Workers other than agricultural workers	21	76	104	1.22	4.55	5.98	2.71	7.79	10.22
5.	Total workers	773	953	107	45.17	57.34	58.37	100.0	100.0	100.0
6.	Non-workers	938	714	721	56.82	42.83	41.48	-	-	-
	Total	1711	1867	1738	100.0	100.0	100.0			

Source : District Census Handbook for 1971, 1981 and Village Panchayat Office, Nagazari.

Resume

The village has basically dominance of agriculture involving 82.64 per cent working force to total workers. Agriculture of village is supported by number of cultivators and agricultural labourers which increased during the span of 14 years. Cotton and hybrid jowar are grown as major crops in kharif crops. Cotton covers largest areal extent of 46.49 per cent to net sown area while hybrid jowar raised on 30.90 per cent to net sown area. Wheat accounts 3.35 per cent to net sown area on silty loam soils. Most of the farmers are still not familiar with improved seeds and fertilizers. The farmers in Nagazari are conservative in their attitude.

3. KACHANGAON

Introduction

Kachangaon lies in the east part in the Yashoda basin, having 1375 hectares areal extent and population of 1265 (1989) with density of 0.96 persons per hectare. It is situated on 78.43' North latitude and 20.20' East longitude on nearly 240 to 250 metres above sea level. The location of this village is given in Exhibit 9.10. It is surrounded by Yarangaon in the east, Pawani in west, Allipur in north and Sirasgaon in South.

The gaothan is at centre within the village area (Exhibit 9.10). The general slope is toward southwest. It is evident from the Nallas flowing to southwest. Kachangaon has 97.07 per cent working force engaged in agricultural practice. Teli and Kunbi constitute the major part of population appear to be involved as cultivators. The scheduled castes and tribes work as agricultural labourers. These people are poor.

Climate and Soils

The village prevails typical climate of monsoon, receiving average annual rainfall between 1000 to 1100 mms. The temperature is similar to Wardha. The intensive heat in summer and fairly cold in winter are common. It is observed that village belongs to medium black soil zone (Exhibit 9.10). The

Table 9.18
Agricultural Operation and Yields

Landuse	*Crops*				
Operations	*Cotton*	*Hybrid Jowar*	*Tur*	*Local Jowar*	*Groundnut*
Botanical Name	Gossypium Spp.	Sorghum Bicolor	Cajanus indicus	Sorghum vulgare	Triticum vulgare
Soil	Clay loam/ sandy clay	Clay loam/ sandy clay	Clay loam/ sandy clay	Clay loam/ sandy clay	Sandy clay
Sowing time	June/July	June	June/July	June	June/July
Spacing	3 to 3 1/2'	7 to 9"	3 to 3 1/2"	9 to 10"	7 to 9"
Seed rate Per acre in Kg.	1 Kg	1 Kg	2 1/2 Kg	1 1/2 Kg	10 to 12 kg
Method of sowing	Dibbling	Drilling	Dibbling	Drilling	Drilling
Manures of Fertilizers (Npk) in Kg. per acre	40-7-0	3-5-0	Nil	Nil	Nil
Irrigation	Rainfed (Kharif)	Rainfed (Kharif)	Rainfed (Kharif)	Rainfed (Kharif)	Rainfed (Kharif)
Harvesting time	4 to 4 1/2 m.	3 months	4 months	3 1/2 months	5 1/2 month
Yield per acre in Kg.	125-175 Kg	125 Kg	100 Kg	75 to 100 Kg	200 Kg

Source : Field work

soil of Kachangaon can be classified into three types, *namely*, sandy loam, clay loam, and clay as depicted in Exhibit 9.10. The clay soil type is confined to western part on 15 per cent on which cotton and tur are successfully raised. Clay loam appears in the extreme southwest and southern parts covering 43.5 per cent of the village area. Cotton, tur and hybrid jowar are cultivated in this soil type. This soil is relatively fertile than sandy loam. Generally clay and clay loam soils can be categorised as deep black soils. The analysis of soil types is given in Table 9.19). Clay loam records the highest pH value of 8.1. The sandy soils restricted to the north and northwest parts of gaothan, with low fertility where hardy crops like mug, mot, local jowar, kultha are grown yielding low per hectare production. This soil possesses 26.70 per cent coarse sand and 29.55 per cent fine sand and has relatively less fertile than former soil types in Kachangaon. The sandy loan covers 42.5 per cent area in Kachangaon.

Table 9.19
Soil Analysis

Soil texture classes	*Soil Types*		
	Silty loam (per cent)	*Clay loam (per cent)*	*Clay (per cent)*
Coarse sand	4.30	8.40	20.76
Fine sand	13.04	11.85	29.55
Silt	20.00	36.25	22.50
Clay	58.00	20.00	18.50
Nitrogen	2.22	0.31	0.26
Phasphate	0.32	Trace	0.84
Polash	33.75	20.00	4.75
pH value	8.00	8.1	8.3

Note : The above soil analysis was carried out in Soil Testing Laboratory, Agriculture Department, Pune-5.

Spatial and Teınporal Analysis of General Landuse

Table 9.20 displays the per cent of land under different categories is shown in Exhibit 9.11. It is observed that village is cultivated four fifth land while forest is almost absence

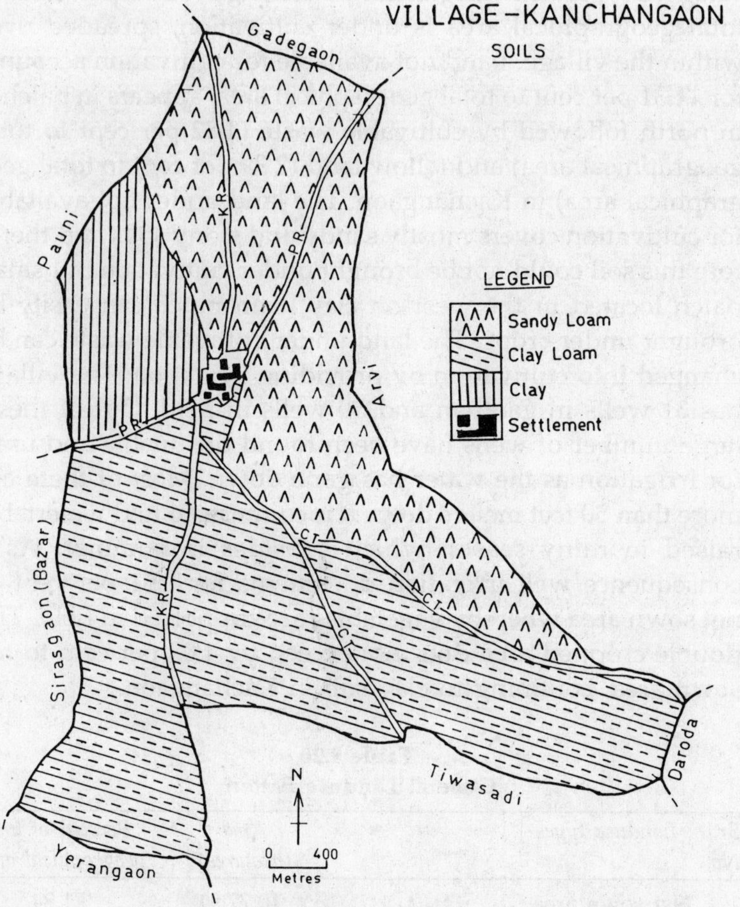
VILLAGE-KANCHANGAON
SOILS
Gadegaon
Pauni
Arvi
Sirasgaon (Bazar)
Yerangaon
Tiwasadi
Daroda
KR
PR
CT
LEGEND
Sandy Loam
Clay Loam
Clay
Settlement
N
0 400
Metres

EXHIBIT — 9.10

except shrubs and disperse trees along Nalla. Other landuse categories present insignificant hectarage. 81.84 per cent to total geographical area is under cultivation, spreaded over within the village. Land not available for cultivation accounts for 11.31 per cent to total geographical area, appears in patches in north followed by cultivable waste (4.82 per cent to total geographical area) and fallow land (2.39 per cent to total geographical area) in Kachangaon. The land under not available for cultivation covers mostly sandy and stony soils and therefore this soil could not be brought under cultivation but small patch located in the west on clay loam patch can easily be brought under crops. The land under cultivable waste can be changed into cultivation by providing irrigation. The village has 31 wells in gaothan and 50 wells in field. Out of these, large number of wells have been found unsuitable and unfit for irrigation as the water has gone down. Most of them are more than 50 feet meters deep. It is common to find watertable raised in rainy season which subsides in summer. As a consequence well irrigation documents for 0.61 per cent to net sown area whereby vegetables are supported in field. The double cropped area does not exceed by 4.03 per cent to net sown area, confining to clay soil in Kachangaon.

Table 9.20
General Landuse Pattern

Sr. No.	*Landuse types*	*Area (hectares)*	*Percent of total geographical area*
1.	Net sown area	10.77.91	81.84
2.	Cultivable waste	59.59	4.52
3.	Fallow Land	30.50	2.33
4.	Land not available for cultivation	149.00	11.31
5.	Forest	Nil	Nil
6.	Double cropped area	53.17	4.03
7.	Total cropped area	1131.08	85.87
8.	Irrigated area	8.14	0.61
9.	Total geographical area	1317.00	—

Source : Village Revenue Records, Kachangaon.

Table 9.20 and Exhibit 9.12 reveal the temporal variations in general landuse pattern from 1975-76 to 1988-89 in Kanchangaon. It is noticed that net sown area is considerably increased during period of study. It has increased by 21.49 per cent to total geographical area in span of fourteen years. The net sown area had been 60.35 per cent in 1975-76 which recorded 61.84 per cent to total geographical area in 1985-86. The increase in percentage under net sown area is visible from 1982-83 onwards (Table 9.21) due to introduction of high yielding varieties of cotton and hybrid jowar. Thereby cultivable waste and fallow land have been reduced and replaced by net sown area. The cultivable waste has reduced to 5.64 per cent followed by fallow land (4.64 per cent to total geographical area). The shrubs and babul trees have been cleared and these land has brought under cultivation. This holds true of land not available for cultivation due to increasing pressure of population on land. The percentage of land not available for cultivation has reduced from 22.40 per cent to 11.31 per cent to total geographical area (1975-76 to 1988-89).

Spatial and Temporal Analysis of Agricultural Landuse

The cropping pattern of Kachangaon is outcome of soil types, slight variations in rainfall amount and traditional approach of agricultural practice. The kharif is the most important season when nearly 70 per cent crops are sown and produced. This season commences from June. Cotton, hybrid jowar, local jowar, tur, mug, mot, groundnut, sesamum are cultivated. Cotton stands as first ranking crop occupying maximum percentage among crops (48.96 per cent to net sown area) in Kachangaon. It is sown on clay and clay loam soils producing high yield per hectare. The spatial distributional pattern of cotton is visible from Exhibit 9.11. It is mainly concentrated in the central and south parts. 21.47 per cent land to net sown area is under hybrid jowar and stands second in hectarage. The cultivation of hybrid jowar is

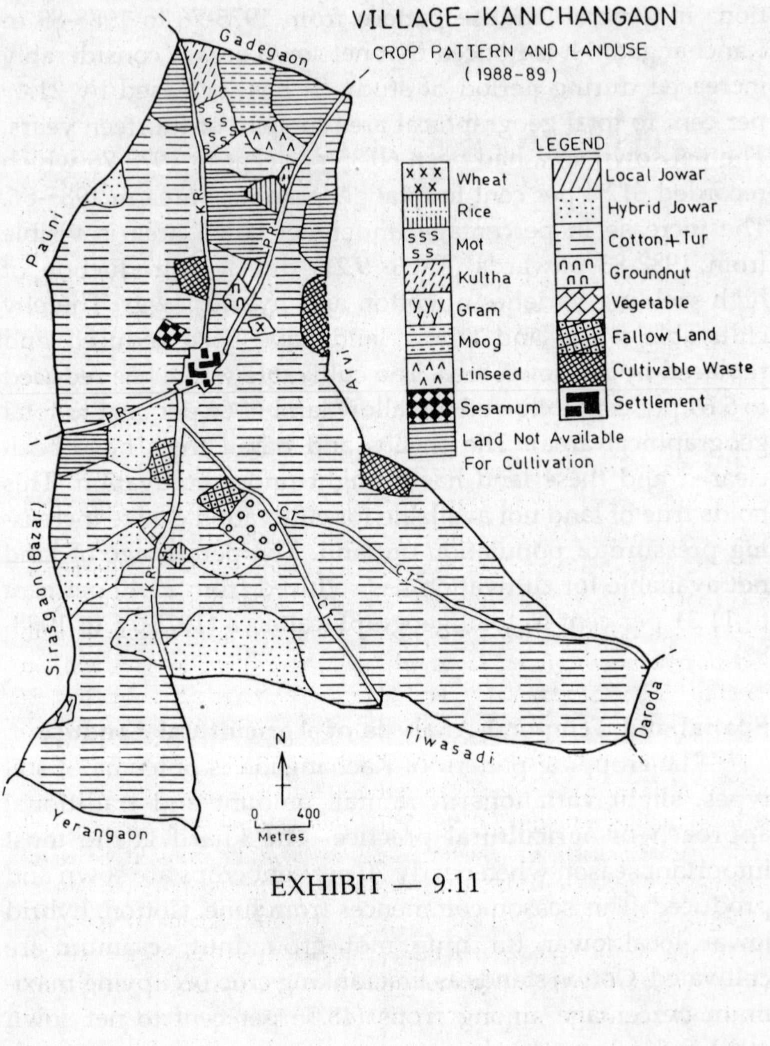
VILLAGE-KANCHANGAON
CROP PATTERN AND LANDUSE
(1988-89)
LEGEND
Wheat
Rice
Mot
Kultha
Gram
Moog
Linseed
Sesamum
Land Not Available For Cullivation
Local Jowar
Hybrid Jowar
Cotton+Tur
Groundnut
Vegetable
Fallow Land
Cultivable Waste
Settlement
Gadegaon
Pauni
Arvi
Sirasgaon (Bazar)
Tiwasadi
Daroda
Yerangaon
K R
P R
C T
N
0 400
Metres

EXHIBIT — 9.11

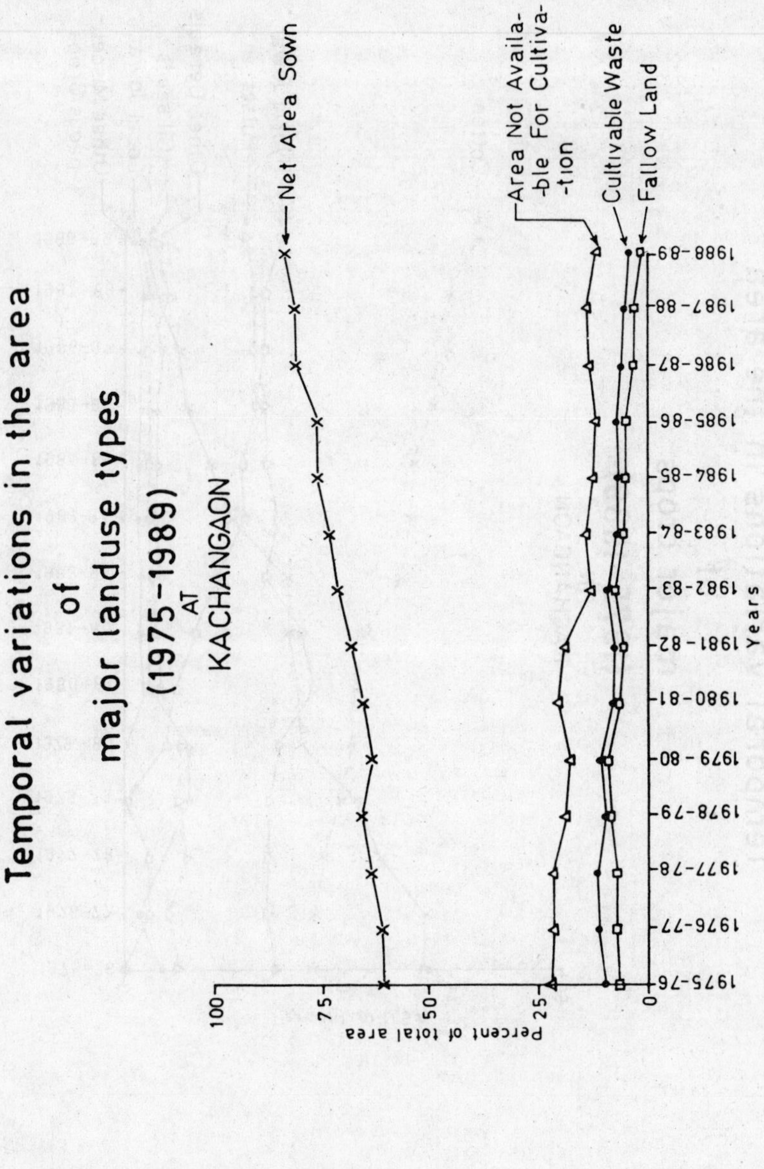
Temporal variations in the area
of
major landuse types
(1975-1989)
AT
KACHANGAON
Net Area Sown
Area Not Availa-
-ble For Cultiva-
-tion
Cultivable Waste
Fallow Land
Percent of total area
100
75
50
25
0
1975-76
1976-77
1977-78
1978-79
1979-80
1980-81
1981-82
1982-83
1983-84
1984-85
1985-86
1986-87
1987-88
1988-89
Years

EXHIBIT — 9.12

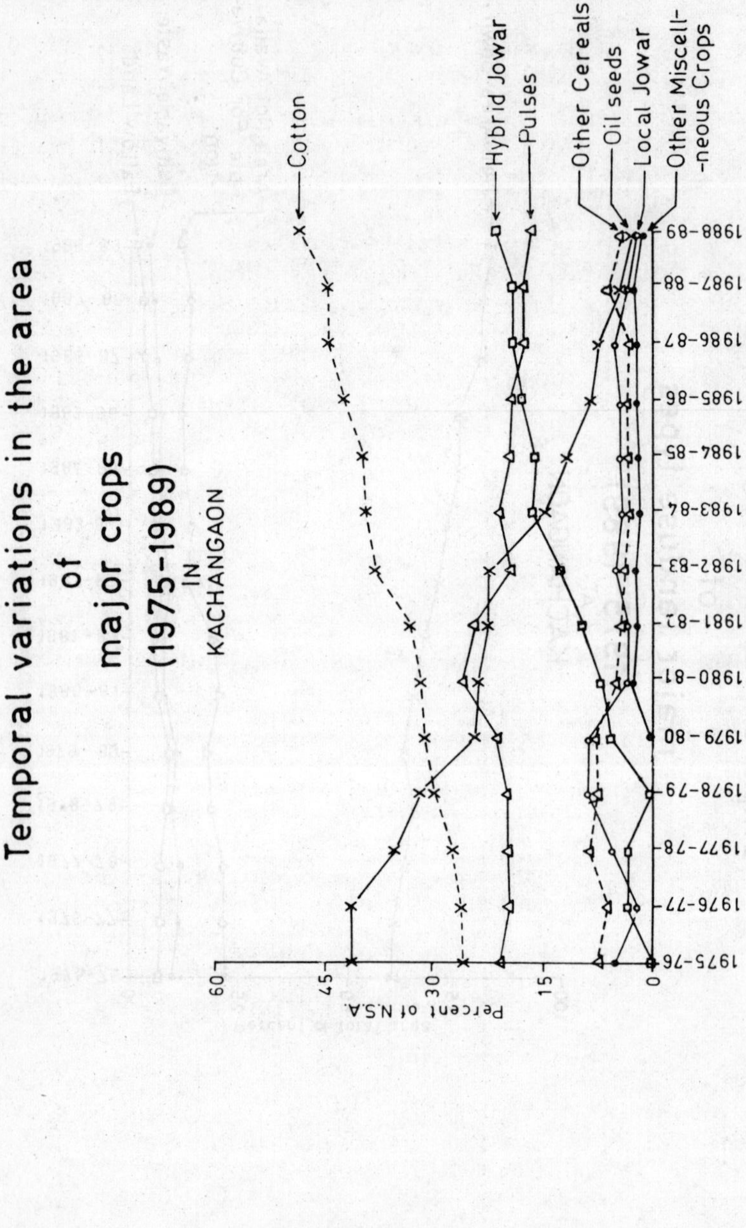
Temporal variations in the area
of
major crops
(1975-1989)
IN
KACHANGAON
Cotton
Hybrid Jowar
Pulses
Other Cereals
Oil seeds
Local Jowar
Other Miscell-
-neous Crops
Percent of N.S.A.
60
45
30
15
0
1975-76
1976-77
1977-78
1978-79
1979-80
1980-81
1981-82
1982-83
1983-84
1984-85
1985-86
1986-87
1987-88
1988-89
Years

EXHIBIT — 9.13

Table 9.21
Temporal Variations in Landuse (1975-76 to 1988-89)

Years	*General Landuse (per cent of total geographical area)*						*Agricultural Landuse (% of net sown area)*					
	NSA	*CW*	*FL*	*ANC*	*Forest*	*Cotton*	*Local Jowar*	*Hybrid Jowar*	*Other cereals*	*Total Pulses*	*Total oil*	*Other miscellaneous seeds*
1975-76	60.35	10.20	7.05	22.40	NIL	25.05	40.38	NIL	7.25	22.00	5.32	NIL
1976-77	60.10	11.27	7.46	21.17	NIL	25.15	40.15	3.72	6.17	20.50	20.20	2.11
1977-78	62.42	10.40	6.18	21.00	NIL	26.30	3.50	8.98	20.11	5.42	NIL	
1978-79	65.51	8.66	7.32	18.51	NIL	30.11	32.51	NIL	7.50	20.50	9.33	NIL
1979-80	63.22	10.33	9.10	17.35	NIL	31.78	25.11	6.19	9.92	21.35	8.65	NIL
1980-81	65.77	7.24	7.88	19.11	NIL	31.59	23.78	7.32	5.53	25.11	3.49	3.58
1981-82	67.39	7.56	6.50	18.55	NIL	34.51	21.33	9.88	4.38	23.52	4.27	2.11
1982-83	70.17	8.87	7.90	13.12	NIL	37.11	20.19	12.57	3.78	19.37	4.48	2.50
1983-84	73.30	6.81	6.30	13.59	NIL	39.55	15.32	15.33	3.50	20.51	3.68	2.11
1984-85	75.15	6.69	5.15	13.00	NIL	39.55	12.00	15.21	3.37	19.88	4.83	5.15
1985-86	76.12	6.93	4.95	12.00	NIL	42.55	9.55	17.39	3.50	19.35	5.14	2.52
1986-87	79.82	4.31	3.57	12.30	NIL	44.11	7.41	19.12	3.27	18.52	5.32	2.25
1987-88	79.32	5.07	3.11	12.50	NIL	45.33	4.50	19.30	4.78	18.10	4.49	3.51
1988-89	81.84	4.52	2.31	11.31	NIL	48.98	1.97	21.47	3.74	16.49	3.71	3.80

Source : Village Revenue Records, Kachangaon

Note : NSA - Net Sown Area, ANC - Area Not Available for Cultivation, FL - Follow Land, CW - Cultivable Land.

observed, in dispersed patches in cotton raising area mainly in the central and south parts on clay and clay loam soils. Tur stands as third major crop in ranking occupying 9.50 per cent to net sown area (Exhibit 9.11) in Kanchangaon. Kultha is sown on 3.48 per cent to net sown area followed by mug (3.34 per cent), wheat (2.90 per cent) and mot (2.88 per cent) distribute in patches within the areal extent of village undertaken for study.

Table 9.22
Agricultural Landuse Pattern

Sr. No.	*Name of crops*	*Percent to net sown area*	*Sr. No.*	*Name of crops*	*Percent to net sown area*
1.	Cotton	48.98	8.	Mot	2.88
2.	Hybrid jowar	21.47	9.	Chilli	0.07
3.	Local jowar	1.97	10.	Vegetables	0.09
4.	Tur	9.50	11.	Wheat	2.90
5.	Sesamum	1.66	12.	Gram	0.77
6.	Mug	3.34	13.	Groundnut	0.20
7.	Rice	0.84	14.	Kultha	3.48
			15.	Linseed	1.85

Source : Village Revenue Records, Kachangaon.

The temporal variations in agricultural landuse patterns are shown in Exhibit 9.13) and Table 9.21. It is visible from Table 9.21 that local jowar had been predominant crop in 1975-76. Now this crop is partly replaced by cotton and partly by hybrid jowar. The overall increase of cotton is 23.93 per cent in Kachangaon during the span of fourteen years. The noteworthy change increasing percentage under cotton after 1981-82 (Exhibit 9.13). Since then it has dominated. The adoptation of new high yield in cotton seeds by farmers have brought such change. That is why notable reduction has found under local jowar (38.41 per cent to net sown area). Moreover, it is observed that with the decrease of local jowar in hectarage there has been increase in cotton. Hybrid jowar has introduced after 1975-76 for the initial stage on 3.72 per cent to net sown area, which registered and attained maximum 21.47 per cent to net sown are in 1985-86. During the period of study, hybrid jowar increased by 17.75 per cent to net sown area. The pulses includes tur, gram and mug. The hectarage under pluses show decline trend in Kachangaon. The hectarage of mug had 22.00 per cent in 1975-78 and it reduced 16.49 per cent to net sown area in 1988-89. Oilseeds also have been reduced by 1.61 per cent to net sown area (Table 9.21 and Exhibit 9.13) during the span of fourteen years.

Landuse and Population

Table 9.23 presents per capita share for different landuse categories in Kachangaon. It is observed that per capita share is 0.14 hectare for village and 0.85 hectare for land available for cultivation. This is comparatively higher than earlier three sample villages showing improvement in agriculture. It was found that area under kharif has 0.80 hectares per capita land whereas it is 0.04 for area under rabi. This explains the significance of agricultural in kharif season than rabi. Double cropped area do not show dominance in the cropping pattern of the village. Table 9.24 set forth scenario of demographic characteristics of Kachangaon. The village has increased population in number during the span of fourteen years. It has registered 1016 persons in 1971 and it is 1265 persons in 1989, with density of 0.77 persons per hectare and 0.96 persons per hectare respectively. The density of population like other villages was found increasing in study period. Still the village has not crossed fifty per cent literacy. This indicate low rate of literacy in Kachangaon, and thus this could be one of the reasons, for not speedy development in agriculture. 97.93 per cent working force to total workers is engaged in agriculture. Cultivators and agricultural labourers both are not increasing in percentage to total workers. Table 9.25 shows decline in cultivators and agricultural labours. The decline of cultivators from 1971 to 1989 is 4.53 per cent to total workers while it is declined by 13.35 per cent to total workers in case of agricultural labourers. The more or less change is found in marginal workers and workers other than four workers in Kachangaon

Table 9.23
Landuse and Population

(*Population 1265*)

Sr. No.	*Landuse categories*	*Area*	*Land per head of population*
1.	Total village area	1317.00	1.04
2.	Land available for cultivation	1077.91	0.85
3.	Total Kharif crops area	1016.11	0.86
4.	Total rabi crops area	61.80	0.04
5.	Cross cropped areas	1131.08	0.89
6.	Double cropped area	53.17	0.04

Source : Author

Table 9.24
Demographic Charateristics

Sr. No.	*Categories*	*Years*		
		1971	*1981*	*1989*
1.	Total Population	1016	1188	1265
2.	Density of Population	0.77	0.90	0.96
3.	Literacy percentage	33.85	41.32	49.80
4.	Per capita land	1.29	1.10	1.04
5.	Percentage of farm workers to total workers	97.71	93.93	97.93

Source : Wardha District Census Handbook for 1971 and 1981, and Village Panchayat Office, Kachangaon.

Agricultural Operation and Yields

The agricultural operation and yields are shown in Table 9.26. Cotton, hybrid jowar, tur, kultha and mug are sown in kharif season in June month. Tur and cotton are sown by dibbling method while mug, hybrid jowar and kultha are sown by drilling method. All these crops are raised in monsoon season, hence are rainfed. The cotton is matured in four to five months. Hybrid jowar is harvested in October. The average yield of cotton and hybrid jowar are 300 Kgs and 200 Kgs per hectare respectively. Tur found low yield per hectare in Kachangaon (75 to 100 Kg per hectare). Kultha is sown in the village on 3.48 per cent to net sown area on inferior soils preferably sandy loam where cotton and hybrid jowar do not get high return. Mug is sown mixed with hybrid jowar. It yields 200 Kgs per hectare in Kachangaon (Table 9.26).

Socio-Economic Facilities and Farmers

It was noticed during the field work and interviews with farmers regarding socio-economic facilities available to farmers in Kachangaon. Soil type is the main aspect reflecting on cropping pattern. The village has more than 50 per cent sandy soils and therefore yield are very low. Poor economic condition limitation of availability of fertilies and pesticides and conservative attitude of society have resulted existing landuse pattern. Three farmers owned more than 10 hectares land, who cultivate land for vegetables besides cotton and hybrid jowar. The vegetables are sold locally limitation of well irrigation. In the village irrigation has not extended area under cultivation at large extent.

Resume

Agriculture is the main occupation of Kachangaon, involving 97.93 per cent working forces to total workers in it. It is noticed that clay loam and clay soils in the south and west are suitable for raising crops *namely*, cotton and hybrid jowar and therefore large areal extent has found under cotton (48.95 per cent to net sown area) and hybrid jowar (21.41 per cent to net sown area). Cotton, hybrid jowar, local jowar, tur, sesamum, rice, mug, mot, kultha groundnut are raised in kharif seasons indicating dominance of agricultural activity. Wheat has found major crop covering 2.90 per cent to net sown area in Kachangaon and stands first in rank in rabi

Table 9.25
Occupational Structure of Kachangaon

Sr.	Categories	Population			Percent of total population			Percent of total population		
		Years								
		1971	*1981*	*1989*	*1971*	*1981*	*1989*	*1971*	*1981*	*1989*
1.	Cultivators	222	263	283	21.85	22.13	22.37	45.96	41.94	41.43
2.	Agricultural labourers	239	136	158	23.52	11.61	12.49	49.48	22.00	23.13
3.	Marginal workers	16	208	222	1.57	17.50	17.54	3.31	33.17	32.50
4.	Workers other than agricultural workers	6	18	20	0.59	1.51	1.58	1.24	2.87	2.92
5.	Total workers	483	627	683	47.53	52.77	53.99	100.0	100.0	100.0
6.	Non-workers	533	561	582	52.46	47.2	-	-	-	-
7.	Total	1016	1188	1265	100.0	100.0	100.0	-	-	-

Source : District Census Handbook for 1971, 1981 and Village Panchayat Office, Kachangaon.

Table 9.26
Agricultural Operation and Yields

Landuse	*Crops*				
Operations	*Cotton*	*Hybrid Jowar*	*Tur*	*Local Jowar*	*Groundnut*
Botanical Name	Gossypium Spp.	Sorghum Bicolor	Cajanus indicus	Sorghum vulgare	Triticum vulgare
Soil	Clay loam/ sandy clay	Clay loam/ sandy clay	Clay loam/ sandy clay	Clay loam/ sandy clay	Sandy clay
Sowing time	June/July	June	June/July	June	June/July
Spacing	3 to 4'	8 to 10"	3 to 4"	5 to 7"	3 to 7"
Seed rateper acre in kg	1 Kg	1 kg	2 1/2 Kg	3 to 3 1/2 Kg	2 to 12 Kg
Method of sowing	Dibbling	Drilling	Dibbling	Drilling	Drilling
Manures of Fertilizers (NPK) in Kg. per acre	40-10-0	15-5-0	Nil	Nil	Nil
Irrigation	Rainfed (Kharif)	Rainfed (Kharif)	Rainfed (Kharif)	Rainfed (Kharif)	Rainfed
Harvesting time	October 200 to 300	September 200 Kg	December 75 to 100 Kg	October 200 Kg	September 200 Kg
Yield per acre in Kg.	125-175 Kg.	125	100-	75 to 100	200 Kg.

Source : Field work

season, grown on clay loam soil (Exhibit 9.11) linseed and gram do not cover extensive land in the village as it gets low yield per hectare to farmers. Well irrigation has not extended in the village as the water table is deep and expensive to dig wells and therefore farmers have not extended irrigation in Kachangaon.

4. KURZADI

Introduction

Kurzadi is located in central part and is representative village of developing agricultural region (Exhibit 9.14). It is 10 Kms away from Wardha. According to 1989 census population is 1102 persons and density of 4.11 persons per hectare while Kurzadi registered 1258 persons in 1989 with density of 4.82 persons per hectare. The village occupies 260.87 hectares are. The gaothan is in south part on Nalla (Exhibit 9.14). The village is confined by Jamnala in north, Asharatpur in east, Banpur and Fattepur in west and Chunhala in south. It is situated on 78° 36' east latitude and 20° 39' North longitude. The general slope of Nalla is towards south (Exhibit 9.14). Teli and Kunbi are cultivators in majority while other tribes are agricultural labourers.

Climate and Soils

Kurzadi experiences monsoon type of climate. Intensive heat in summer and intensive cold in winter are common to the village. The rainfall received from southwest monsoon prove to be useful for varieties of crops. The personal observations during field work and soil analysis uncover three soil types in the villages (Exhibit 9.14), *namely*, clay, clay loam, and sandy clay soils. The analysis of soil types of Kurzadi is given in Table 9.27. Sandy soil is found on extensive area in north, northwest and southeast parts in Kurzadi covering 40 per cent to total village area. It contains coarse sand and fine sand, 15.65 and 33.03 per cent respectively. It is relatively low fertile than other two soil types. Mug and mot can be yielded successfully. Clay soil confines to west and northwest part containing 46.50 per cent clay (Table 9.27 and Exhibit 9.14) while clay loam is observed in west and southwest parts, west of gaothan covering only 23 per cent area of Kurzadi. It obtains 38 per cent clay and 29.75 per cent silt.

Table 9.27
Soil Analysis

Soil texture classes	*Soil Types*		
	Silty loam (per cent)	*Clay loam (per cent)*	*Clay (per cent)*
Coarse sand	19.76	17.30	15.65
Fine sand	14.96	9.70	33.03
Silt	12.50	28.75	6.25
Clay	46.50	38.00	41.75
Nitrogen	0.73	0.26	0.29
Phasphate	0.80	0.84	1.36
Polash	37.56	52.50	41.25
pH value	7.9	7.8	8.0

Note : The above soil analysis was carried out in Soil Testing Laboratory, Agriculture Department, Pune-5.

Spatial and Temporal Analysis of General Landuse

The general landuse pattern of Kurzadi is given in Table 9.28. Kurzadi has 205.36 hectare accounting 78.72 per cent to total geographical area under cultivation. Forest is almost absences except disperse shurbs and trees along Nalla. The cultivated land in Kurzadi is well spread over the village area, with major concentration in fertile soils. Fallow land and land not available for cultivation account for 7.66 per cent and 7.86 per cent to total geographical area respectively. Five patches of fallow land have been observed in Kurzadi. Major extensive patch of fallow land lies in northwest part while two patches in northeast part along the border of Ashartpur. The patch located in northwest is confined to high relief. Cultivable waste appears in three patches located in northeast and west. Largest patch of

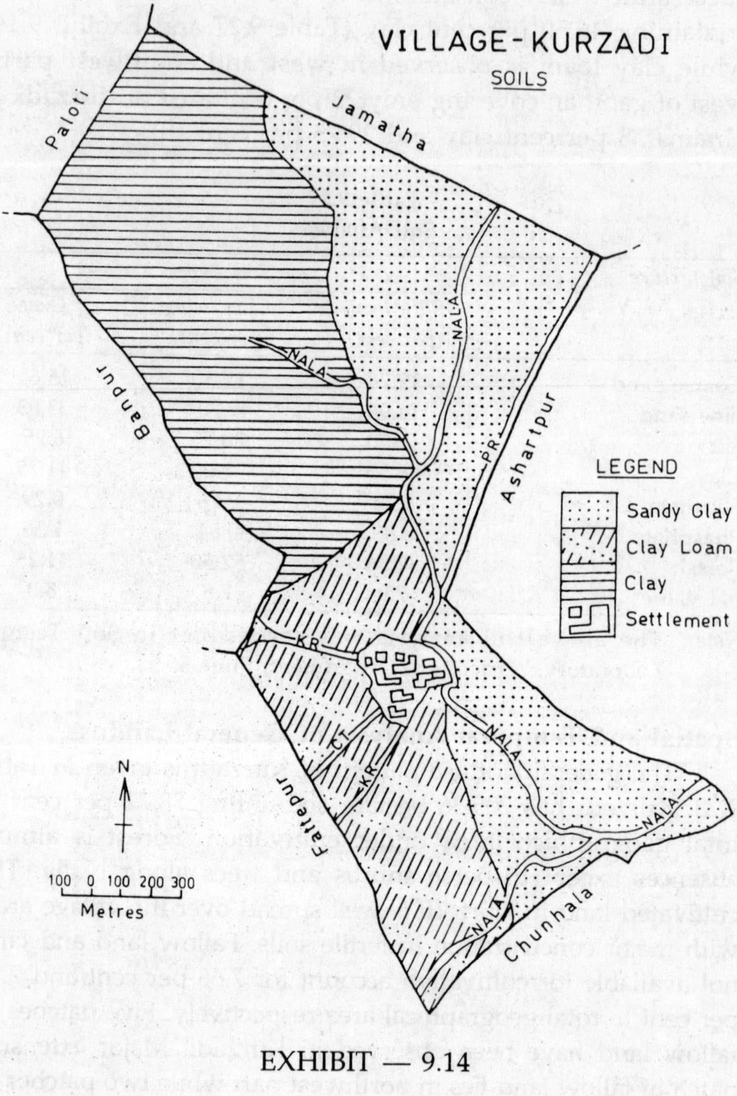
VILLAGE-KURZADI
SOILS
Paloti
Jamatha
NALA
NALA
Banpur
PR
Ashartpur
LEGEND
Sandy Glay
Clay Loam
Clay
Settlement
KR
CT
KR
Fattepur
NALA
NALA
NALA
Chunhala
N
100 0 100 200 300
Metres

EXHIBIT — 9.14

areal extent of cultivable waste is found in northeast in east of Nalla. The land under this category could be brought into cultivation by proper planning. Western patch was kept under cultivable waste in order to regain fertility (Interview with farmers). Kurzadi has 24 wells in field and 7 wells in gaothan. The wells near villages are used for irrigation on 10.25 hectares (13.92 per cent net sown area). Wheat, rice and vegetables are raised on irrigated land. Double cropping accounts for 5.82 per cent total geographical area. Large size holders are found raising double crops in their fields in rabi season.

Table 9.28
General Landuse (1988-89)

Sr. No.	*Landuse types*	*Area (hectares)*	*Percent of total Geographical area*
1.	Net sown area	205.36	78.72
2.	Cultivable waste	15.00	5.74
3.	Fallow Land	20.00	7.66
4.	Land not available for cultivation	20.51	7.86
5.	Forest	Nil	Nil
6.	Total cropped area	220.56	84.54
7.	Irrigated area	10.25	3.92
9.	Total Geographical area	260.87	Nil

Source : Village Revenue Records, Kurzadi.

Table 9.29 and Exhibit 9.16 present the temporal variations in general landuse pattern from 1975-76 to 1988-89 in Kurzadi. It is noticed from table 9.6 that net sown area in the village shows steady upward trend during the study period. It has 68.10 per cent to total geographical area in 1975-96 while it is 78.72 per cent to total geographical area in 1988-89. The overall increase in net sown area is 10.62 per cent to total geographical area. The increase under net sown area is observed from 1975-76 (Exhibit 9.16) and Table 9.29. Among other categories of landuse types have shown more or less decline trend in Kurzadi. All these landuse types are declined not more than 8 per cent to total geographical area. The highest decline found in case of land not available (8.27 per cent to

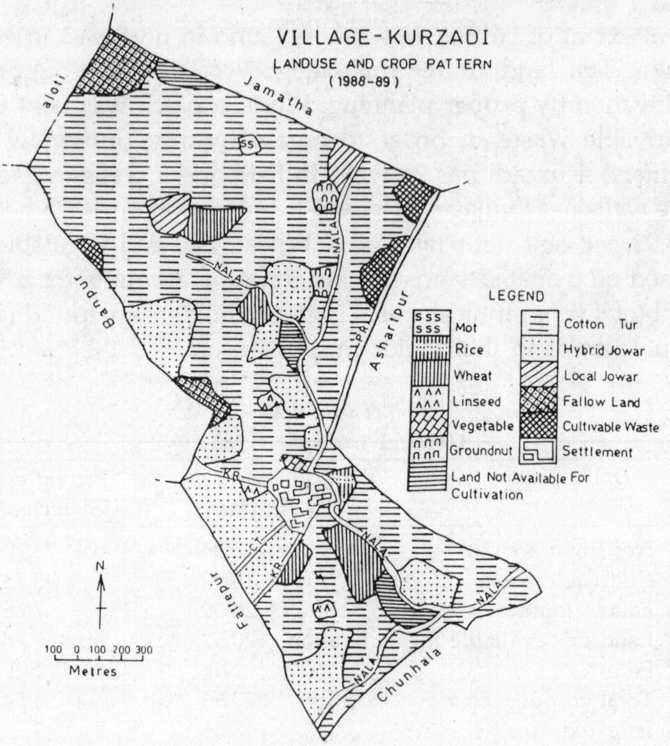
VILLAGE-KURZADI
LANDUSE AND CROP PATTERN
(1988-89)
Jamatha
aloti
Banpur
Ashartpur
Fattepur
Chunhala
NALA
PR
KR
CT
LEGEND
Mot
Rice
Wheat
Linseed
Vegetable
Groundnut
Land Not Available For Cultivation
Cotton Tur
Hybrid Jowar
Local Jowar
Fallow Land
Cultivable Waste
Settlement
N
100 0 100 200 300
Metres

EXHIBIT — 9.15

Temporal variations in the area of major landuse types (1975–1989) AT KURZADI

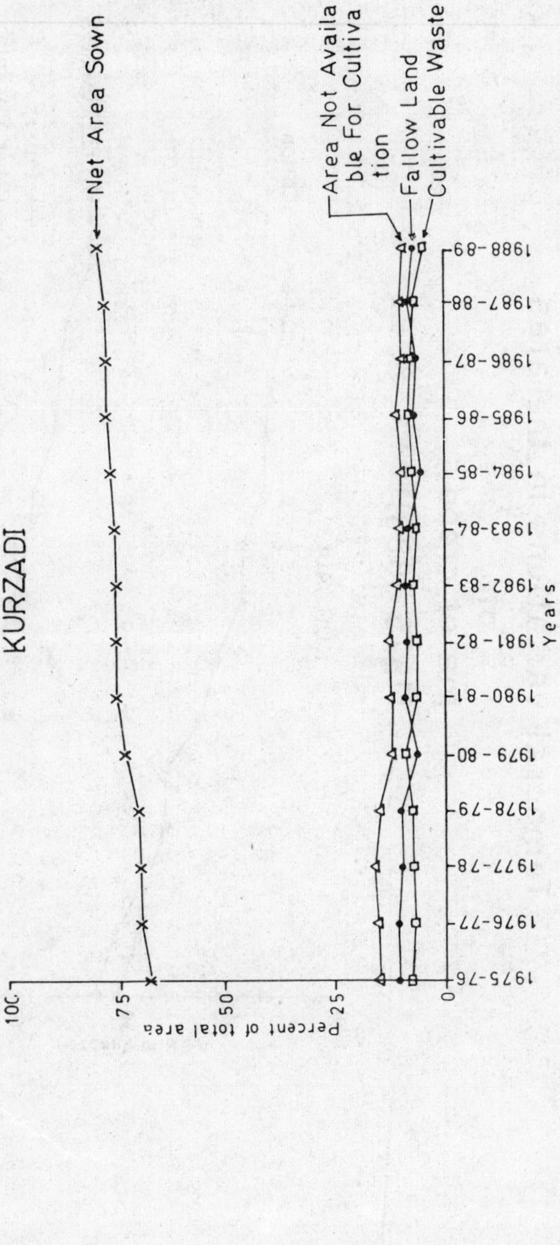

EXHIBIT — 9.16

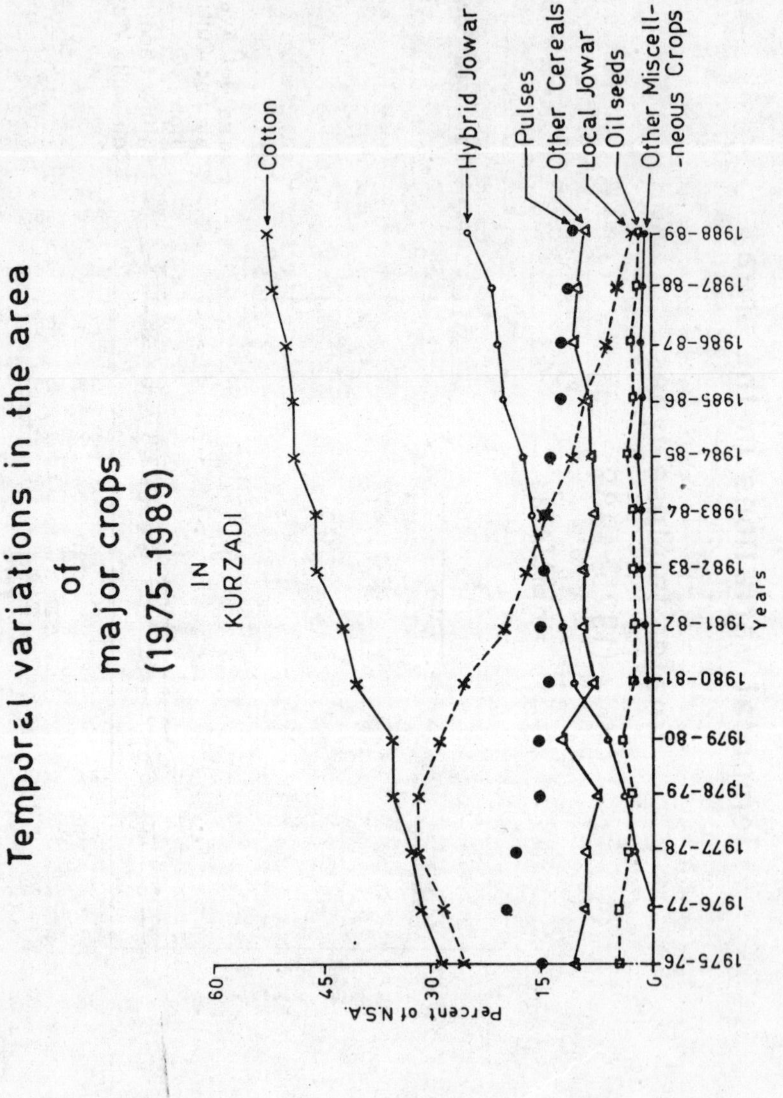
Temporal variations in the area
of
major crops
(1975-1989)
IN
KURZADI
Percent of N.S.A.
60
45
30
15
0
1975-76
1976-77
1977-78
1978-79
1979-80
1980-81
1981-82
1982-83
1983-84
1984-85
1985-86
1986-87
1987-88
1988-89
Years
Cotton
Hybrid Jowar
Pulses
Other Cereals
Local Jowar
Oil seeds
Other Miscell-
-neous Crops

EXHIBIT — 9.17

Table 9.29

Temporal Variations in Landuse (1975-76 to 1988-89)

Years	*General Landuse (percent of total geographical area)*					*Agricultural Landuse (% of net sown area)*						
	NSA	*CW*	*FL*	*ANC*	*Forest*	*Cotton*	*Local Jowar*	*Hybrid Jowar*	*Other cereals*	*Total Pulses*	*Total oil*	*Other miscella-neous seeds*
1975-76	68.10	6.62	10.15	15.13	Nil	28.25	33.11	Nill	10.15	22.37	3.62	2.50
1976-77	69.55	6.06	9.37	15.02	Nil	31.31	35.32	Nil	8.60	20.16	4.61	Nil
1977-78	69.20	7.24	8.45	15.11	Nil	33.43	34.10	2.10	9.35	18.25	2.77	Nil
1978-79	70.50	6.79	8.39	14.32	Nil	35.19	31.55	3.59	8.17	15.30	2.62	3.78
1979-80	73.50	8.11	7.80	10.59	Nil	35.11	28.62	5.75	11.52	15.18	3.82	Nil
1980-81	75.17	6.05	8.10	10.68	Nil	39.83	25.32	10.15	8.10	13.10	1.90	1.60
1981-82	75.30	6.24	7.92	10.54	Nil	42.25	14.30	15.38	8.69	13.81	3.15	Nil
1982-83	75.27	6.95	8.20	9.58	Nil	44.58	16.27	14.18	9.23	13.67	2.14	Nil
1983-84	75.50	7.41	7.50	9.53	Nil	45.62	14.30	15.38	7.56	13.81	3.15	1.18
1984-85	76.67	7.17	6.35	9.81	Nil	48.21	10.82	16.52	7.80	13.22	1.40	2.35
1985-86	76.51	7.38	7.39	8.72	Nil	48.30	8.53	19.20	8.	11.30	1.92	1.80
1986-87	76.30	8.28	7.11	8.31	Nil	49.11	5.41	9.65	9.65	11.59	2.32	1.58
1987-88	76.78	7.83	7.50	7.89	Nil	50.27	4.35	21.15	10.31	10.77	1.95	1.20
1988-89	78.72	5.74	7.66	7.86	Nil	51.61	2.51	24.34	7.88	9.73	2.92	0.97

Source : Village Revenue Records, Kurzadi

Note : NSA - Net Sown Area, ANC - Area Not Available for Cultivation, FL - Follow Land, CW- Cultivable Land.

total geographical area) followed by fallow land (2.57 per cent to total geographical area) and cultivable waste (0.88 per cent to total geographical area). The land not available for cultivation has comparatively more decline in per cent to rugged in Kurzadi.

Spatial and Temporal Analysis of Agricultural Landuse

The cropping pattern of Kurzadi is the result of physio-socio-economic conditions prevalent in the village. Two seasons are noticed one Kharif and other rabi. Kharif season is dominant as nearly 80 per cent crops are sown in this season. The crops in kharif are sown in June and July depending on commencing southwest monsoon. Cotton, tur, mug, mot, rice, local jowar, and hybrid jowar are grown in this season. Among these crops, cotton stands first occupying 51.61 per cent land to net sown area. It is generally cultivated on clay loam and clay soils in the west, north and south in Kurzadi. Clay loam soils give comparatively high yield per hectare than clay and sandy loam and therefore cotton in Kurzadi is raised mainly on clay loam (Exhibit 9.15). Hybrid jowar occupies 24.34 per cent land to net sown area in south and central on clay and clay loam soils. Its distribution is found in many patches disintegrated within the village area. It is visible from Exhibit 9.15. Tur is sown mixed with cotton on 8.27 per cent land. Local jowar accounts for 2.15 per cent to net sown area in north part on sandy clay soils. Rabi season commences from October when wheat, gram and linseed are cultivated. Among rabi crops, wheat is major covering 7.38 per cent land to net sown area. The distribution of wheat is observed in four disintegrated patches. The largest patch in areal extent of wheat is found in west of Nall in south part on clay loam soils. Other patches are found in north and west parts. Linseed and gram have not occupied larger area in rabi season. Linseed accounts for 1.46 and gram 1.22 per cent to net sown area.

Table 9.30
Agricultural Landuse Pattern

Sr. No.	*Name of crops*	*Percent to net sown area*	*Sr. No.*	*Name of crops*	*Percent to net sown area*
1.	Cotton	51.61	7.	Vegetables	0.97
2.	Hybrid jowar	24.34	8.	Wheat	7.30
3.	Local jowar	2.51	9.	Hot	1.46
4.	Tur	8.27	10.	Linseed	1.46
5.	Rice	0.58	11.	Groundnut	1.46
6.	Vegetables	0.97			

Source : Village Revenue Records, Kurzadi.

The temporal variation in major crop landuse patterns are shown in Table 9.29 and Exhibit 9.17. It is clear from Exhibit 9.9 that cotton hectarage had a steadily upward trend after 1978-79. The overall increase of cotton in Kurzadi is 22.36 per cent to net sown area during the span of fourteen years. It had 28.25 per cent to net sown area in 1975-96 while it increased to 1988-89 in 51.61 per cent to net sown area. This fact explains the continuous domination of single cotton crop in Kurzadi. Hybrid jowar also shows upward trend of total increase of 22.24 per cent to net sown area during study period. The varieties of hybrid jowar was introduced in 1977-78. It was further observed that the rate of increase of hybrid jowar is less than cotton. Pulses, other cereal, local jowar, oilseeds and other miscellaneous crops have not increased in Kurzadi. The area under these crops are mostly related to attitude of local farmers. The highest decline has observed in case of pulses. It declined by 22.64 per cent to net sown area. Pulses had been dominating hectarage in 1975-76 occupying 22.37 per cent land to net sown area while it is 9.73 per cent of net sown area in 1988-89 in the villages. The decline of local jowar in Kurzadi due to introduction of high yield varieties of hybrid jowar. It declined by 30.60 per cent followed by other miscellaneous crops (1.53 per cent) and other cereal (2.2) per cent to net sown area. This decline in per cent under crops namely, pulses, local jowar, other cereals and oilseeds are noticeable in Exhibit 9.17 during the span of fourteen years in Kurzadi.

Landuse and Population

Table 9.31 summarised the per capita share of land to various categories of landuse in Kurzadi. Per capita land is 0.20 hectare for the village. It was further reduced to kharif and rabi crops area. The area under kharif has 0.14 hectare per capita land while in rabi it is 0.02 hectare. This fact explains the dominance of agricultural activity in kharif season. Double cropping is insignificant hence it shows on 0.01 hectare per capital land in the village. Like other villages, population of Kurzadi shows upward trend of population from 1971 to 1981. The population in number increased during span of fourteen years is 208. The density of population is 4.82 persons per hectare. 85.04 per cent workers are engaged in agriculture, both cultivators and agricultural labourers. Literacy is increased by 13.37 per cent during study period. This indicates slight awareness among the villagers. Agricultural labourers have increased in Kurzadi. It had 46.91 per cent to total workers in 1971 while it is 57.82 per cent to total workers in 1989. Cultivators have found decline trend in Kurzadi. It declined by 13.72 per cent to total workers attributed to migration to Wardha City. The trend of marginal workers and workers other than agriculture are more or less static in the village (Table 9.33).

Table 9.31
Landuse and Population

Sr. No.	*Landuse categories*	*Area (hactare)*	*Land per head of population*
1.	Total village area	260.87	0.20
2.	Land available for cultivation	205.36	0.16
3.	Total Kharif crops area	184.36	0.14
4.	Total rabi crops area	21.00	0.02
5.	Cross cropped areas	220.56	0.17
6.	Double cropped area	15.20	0.01

Source : Author

Table 9.32
Demographic Characteristics

Sr. No.	Categories	Years 1971	1981	1989
1.	Total Population	1050	1102	1258
2.	Density of Population	4.02	4.41	4.82
3.	Literacy percentage	38.85	48.61	52.22
4.	Per capita land	0.24	0.22	0.20
5.	Percentage of farm workers to total workers	89.81	90.46	85.04

Source : Wardha District census Handbook and Village Panchayat Office, Kurzadi.

Agricultural Operation and Yields

The agricultural operation and yields are shown in Table 9.34 cotton, hybrid jowar, tur, and local jowar are kharif crops and are sown in June. The cotton and hybrid jowar can get high yield on clay and clay loam while other crops are grown in all soil types. Drilling method is common for sowing. Comparatively cotton gets high yield per hectare (300 to 400 Kgs) in Kurzadi. Local jowar is also raised in Kurzadi in kharif season but hectarage under this crop has been reduced remarkably due to introduction of high yield varieties of hybrid jowar in place of local jowar. Tur is rabi crop grown in October in moisture retaining soils. Cotton yield mainly depends on soil types, application of fertilizers and timely sprayings of pesticides. Local jowar and wheat generally are not sprayed.

Socio-Economic Facilities and Farmers

Personal interviews with farmers unfold socio-economic facilities available in Kurzadi. The overall situation of Kurzadi is same as earlier villages. Large size holders afford improved seeds, implements in agricultural practice while small land holders can not afford because they are poor economically. The little vegetables are raised to send to Wardha market but has no influence on agricultural pattern. Agricultural

Table 9.33

Occupational Structure of Kurzadi

Sr. No.	Categories	*Population*			*Percent of total population*			*Percent of total population*		
		Years								
		1971	*1981*	*1989*	*1971*	*1981*	*1989*	*1971*	*1981*	*1989*
1.	Cultivators	133	180	193	12.66	15.62	15.34	41.04	28.51	27.22
2.	Agricultural	152	389	410	14.47	33.76	32.59	46.91	33.76	57.62
3.	Marginal workers	14	1	14	1.33	00.08	1.97	4.32	00.15	1.97
4.	Workers other than agricultural workers	25	60	92	2.38	5.20	12.97	7.71	9.52	13.68
5.	Total workers	324	630	709	30.85	54.68	56.35	100.0	100.0	100.0
6.	Non-workers	726	522	549	69.14	45.31	43.64	-	-	-
7.	Total	1050	1152	1258	100.0	100.0	100.0	-	-	-

Source : Wardha District Census Handbook for 1971, 1981 and Village Panchayat Office, Kurzadi.

commodities are sold in local market. Transport facilities are limited. Well irrigation facilities are available in Kurzadi. Size of land holding varies from farmers to farmers. 75 per cent land is observed less than 2 hectares followed by 15 per cent to total land holding under 2 to 4 hectares and 10 per cent to total land holdings are owned by big farmers.

Resume

Kurzadi has dominant agricultural activity involving 85.04 per cent working force to total workers in agriculture. It is observed that high yielding crops are raised on fertile soils while low yielding crops are cultivated on sandy loam soils. Cotton and hybrid jowar are dominant crops and raised on 51.61 per cent 24.34 per cent to net sown area respectively on clay loam and clay soils in Kurzadi. Among other crops grown in kharif, tur covers 8.27 per cent to net sown area. It is sown mixed with cotton. Local jowar does not indicate significantly after 1975 as it is evident from Table 9.29 that areal extent of local jowar has been reduced and replaced by hybrid jowar. It is noticed that wheat is major crop in rabi having raised on 7.30 per cent to net sown area on clay soil in Kurzadi. Other crops do not show any significance in the cropping pattern. Comparatively this village has irrigated area of 3.92 per cent to net sown area indicating positive change in the attitude of farmers in improving agricultural pattern. Further it is noticed that irrigation facilities can be extended by digging wells which could be useful to cultivate crops like rice, wheat and vegetables. This indicates the potentiality to develop agriculture of the village. It is, therefore, this village can be a representative of developing village in the basin.

5. ROHANKHEDA

Introduction

Rohankheda is located in south part in the Yashoda basin (Exhibit 9.1) with an areal extent of 572.26 hectare. It is 25 kms away from Wardha and 30 kms from Hinganghat.

Table 9.34

Agricultural Operation and Yields

Landuse	*Crops*				
Operations	*Cotton*	*Hybrid Jowar*	*Tur*	*Local Jowar*	*Groundnut*
Botanical Name	Gossypium Spp.	Sorghum Bicolor	Cajanus indicus	Sorghum vulgare	Triticum vulgare
	Clay loam/ sandy clay	Clay loam/ sandy clay	Clay loam/ sandy clay	Clay loam/ sandy clay	Sandy clay
Sowing time	June/July	June	June/July	June	June/July
Spacing	3 to 4'	7 to 10"	3 to 4"	9 to 7"	7 to 7"
Seed rate Per acre in Kg.	1 Kg	1 Kg	2 1/2 kg	3 to 3 1/2 Kg	2 to 12 Kg
Method of sowing	Dibbling	Drilling	Dibbling	Drilling	Drilling
Manures of Fertilizers (NPK) in Kg. per acre	40-10-0	15-5-0	Nil	Nil	Nil
Irrigation	Rainfed (Kharif)	Rainfed (Kharif)	Rainfed (Kharif)	Rainfed (Kharif)	Rainfed (Kharif)
Harvesting time	October 200 to 300	September 200 Kg	December 75 to 100 Kg.	October 200 Kg.	September 200 Kg.
Yield per acre in Kg.	125-175 Kg.	125	100-	75 to 100 Kg.	200 Kg

Source : Village Revenue Records, Kurzadi.

According to 1991 census the population is 662 persons and 760 persons in 1961 with density of 1.15 persons per hectare and 1.36 persons per hectare respectively. The village is situated on 78°.38" East latitude and 20°.32" North longitude. The general slope of village is towards east which is evident from the flow of water of Nalla (Exhibit 9.1). The surrounding villages are noticeable from Exhibit 9.18. It is confined by Chanki in south, Gadegaon and Alamdoh in east, Takali in north and Mendukdoh in west. The village is linked by kachha road to Kangaon (5 kms), Giroli (7 kms), Chanaki (2 kms) and Gadegaon (5 kms). The gaothan is located in the centre of village area.

Climate and Soil

Monsoon climate prevails in Rohankheda experiencing intensive heat in summer and fairly cold in winter. Rainfall in monsoon proves of immense importance between June to October for proper cultivation and growth of crops. The soil of Rohankheda is of medium black with minute variations in texture and colour. There are two soil types, namely, clay loam and clay. It is clear from Exhibit 9.18 that these soil types have divided the village into two divisions from gaothan. There is extensive coverage of clay soil in north while clay loam lies in south. Clay loam covers 62 per cent areal extent with high percentage of clay *i.e.* 43 per cent (Table 9.34). This soil type is fertile than that of clay in the village where cotton and hybrid jowar are grown successfully. Other soil type 'clay' is confined to south part on 38 per cent area to village area having 23.75 per cent silt and 35.50 per cent clay and pH value is of 8.1.

Spatial and Temporal Analysis of General Landuse

Out of 572.26 hectares, 532.10 hectares land is under crops accounting 92.98 per cent to the total geographical area. It spreads all over within the village. The large areal extent under cultivation is due to nearly plain part with fertile soil except few small patches of other categories. Other landuse types in Rohankheda are observed in patches located disperse within the village area. The land not available for cultivation accounts for 5.60 per cent to total geographical area in southwest and north west parts. Fallow land is found in west part on the border area of Rohankheda occupying 0.65 per cent to total geographical area whereas cultivable land appears on the left side of kachcha road which leads to Babpur (Exhibit 9.19), accounting for 0.69 per cent to total geographical area. Comparatively Rohankheda has little land under the heads of cultivable waste (0.69 per cent) and fallow land (0.65 per cent). This is due to extensive fertile tract that has been brought under cultivation by farmers in the village. Irrigation is confined to north part in two patches along Nalla on 1.56 per cent to net sown area where rice and vegetables are grown for local consumption.

Table 9.34
Soil Analysis

Soil Texture Classes	*Sandy clay (%)*	*Clay loam (%)*
Coarse sand	12.90	26.66
Fine sand	24.40	5.42
Silt	23.75	20.00
Clay	35.50	43.00
Nitrogen	0.35	0.35
Phosphate	0.80	0.84
Potash	12.75	12.75
pH value	8.1	8.5

Note : The above analysis was carried out in Soil Testing Laboratory, Department of Agriculture, Pune 411005.

Exhibit 9.20 and Table 9.36 display the temporal changes in different landuse types in Rohankheda during the span of fourteen years. It is observed that the land under cultivation (NSA) increased to 24.38 per cent to total geographical area. It had 68.50 per cent to total geographical area in 1975-78

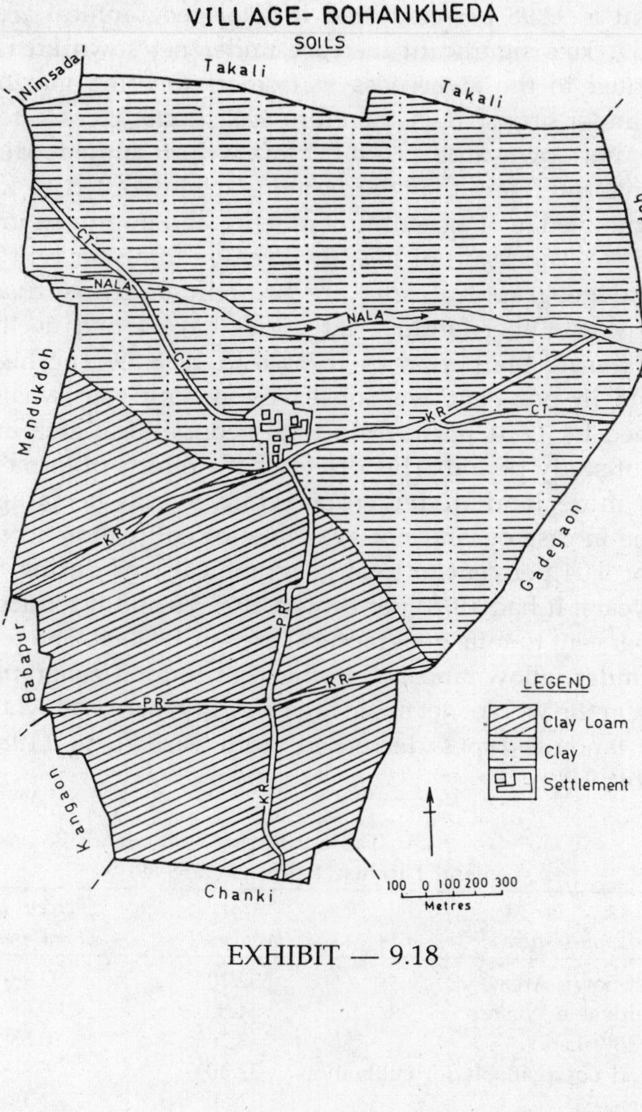
VILLAGE-ROHANKHEDA
SOILS
Nimsada
Takali
Takali
Alamdoh
Mendukdoh
NALA
NALA
CT
CT
CT
KR
KR
KR
KR
PR
PR
Gadegaon
Babapur
Kangaon
Chanki
LEGEND
Clay Loam
Clay
Settlement
100 0 100 200 300
Metres

EXHIBIT — 9.18

while it is 92.98 per cent land to total geographical area in 1988-89. This significant increase under net sown area was attributed to the awareness of farmers to bring maximum land under crops. It is therefore vast hectarage are found under net sown area (92-98). Moreover, it unfolds during personal interview that the farmers are interested to know new knowledge of agricultural practice and improved implements etc. The result of this, the average three per cent land to total geographical area are brought under cultivation every year. Other landuse categories have found declining trend during the period of 1975-76 to 1988-89. The highest decline in per cent was observed in cultivale waste. It declined by 12.31 per cent to total geographical area in the span of study period. This discloses the efforts of farmers to bring more land under crops while lowest hectarage is decline in case of land not available for cultivation accounting for 5.04 per cent to total geographical area during fourteen years. It had 10.70 per cent in 1975-76 and it reduced to 5.60 per cent to total geographical area in 1988-89. The reduction under fallow land is 7.15 per cent to total geographical area. On the whole, net sown area steadily increased whereas other landaus types decreased from 1975-76 to 11988-89 (Exhibit 9.20).

Table 9.35
General Landuse Patterns (1988-89)

Sr. No.	*Landuse types*	*Area (hectares)*	*Percent of total area*
1.	Net Sown Area	532.10	92.98
2.	Cultivable Waste	4.00	0.69
3.	Fallow Land	3.76	0.85
4.	Land not available for cultivation	32.40	5.66
5.	Forest	Nil	Nil
6.	Double Cropped Area	23.00	4.18
7.	Total Cropped Area	556.08	97.6
8.	Irrigated Area	8.98	1.56
9.	Total Geographical Area	572.26	—

Source : Village Revenue Record, Rohankheda.

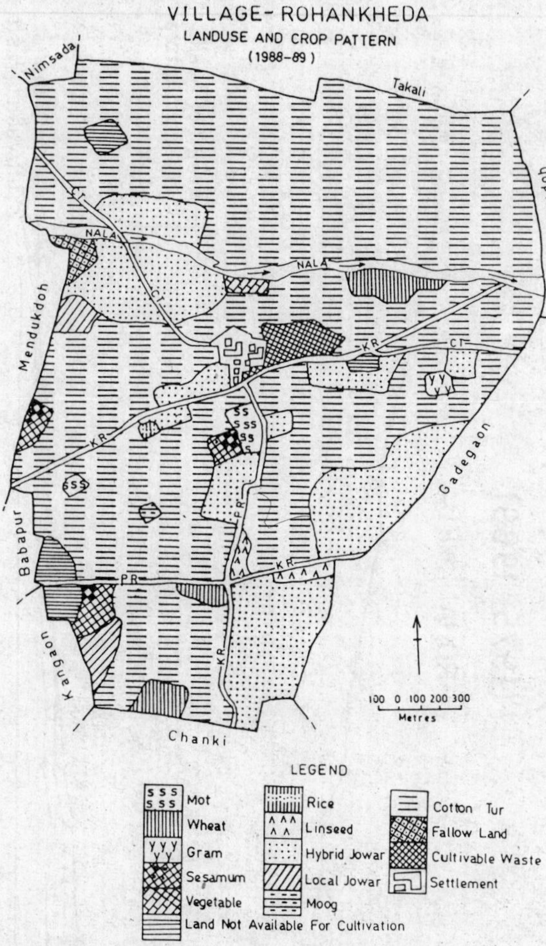
VILLAGE - ROHANKHEDA
LANDUSE AND CROP PATTERN
(1988-89)
Nimsada
Takali
Alamdoh
Mendukdoh
Gadegaon
Babapur
Kangaon
Chanki
NALA
CT
KR
PR
100 0 100 200 300
Metres
LEGEND
Mot
Wheat
Gram
Sesamum
Vegetable
Rice
Linseed
Hybrid Jowar
Local Jowar
Moog
Cotton Tur
Fallow Land
Cultivable Waste
Settlement
Land Not Available For Cultivation

EXHIBIT — 9.19

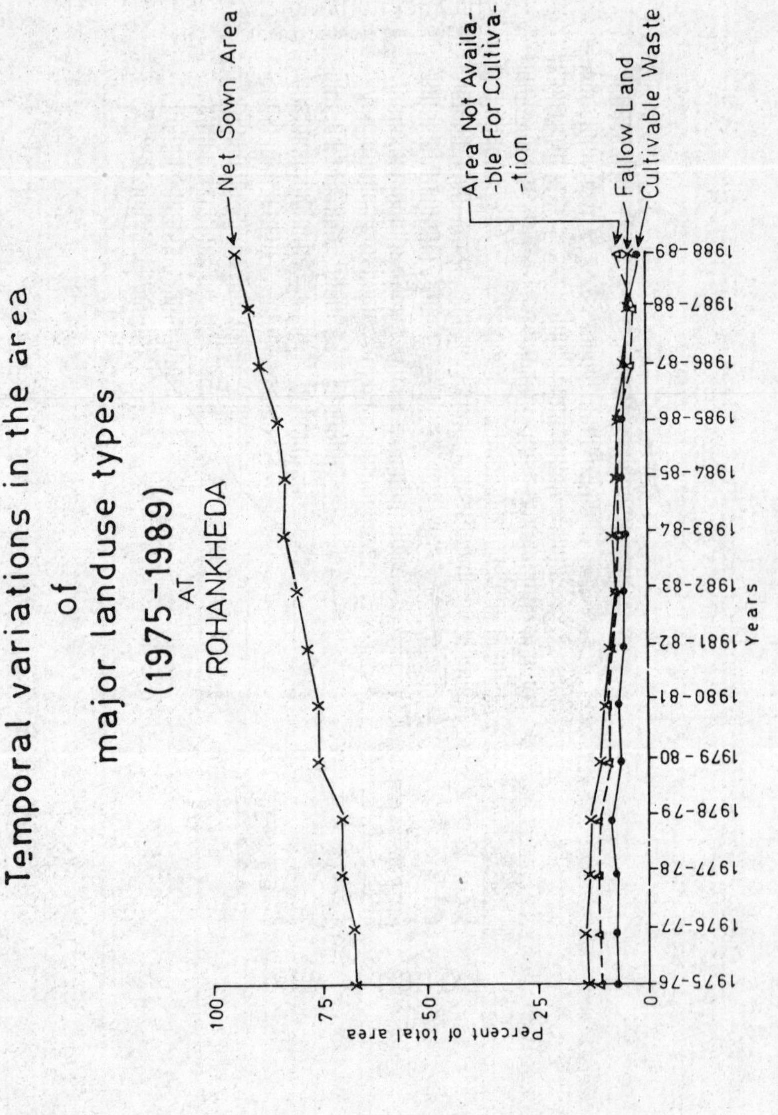
Temporal variations in the area
of
major landuse types
(1975–1989)
AT
ROHANKHEDA
Net Sown Area
Area Not Availa-
-ble For Cultiva-
-tion
Fallow Land
Cultivable Waste
Percent of total area
100
75
50
25
0
1975-76
1976-77
1977-78
1978-79
1979-80
1980-81
1981-82
1982-83
1983-84
1984-85
1985-86
1986-87
1987-88
1988-89
Years

EXHIBIT — 9.20

Temporal variations in the area of major crops (1975-1989) IN ROHANKHEDA

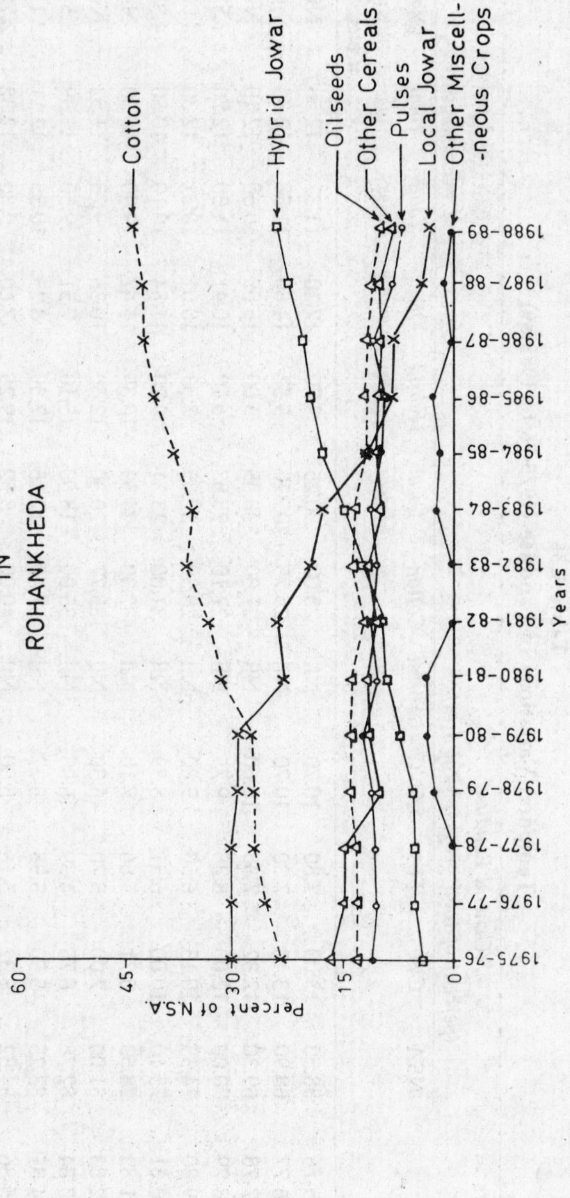

EXHIBIT — 9.21

Table 9.36
Temporal Variations in Landuse (1975-76 to 1988-89)

Years	*General Landuse (percent of total geographical area)*								*Agricultural Landuse (% of net sown area)*			
–	*NSA*	*CW*	*FL*	*ANC*	*Forest*	*Cotton*	*Local Jowar*	*Hybrid Jowar*	*Other cereals*	*Total Pulses*	*Total oil*	*Other miscella-neous seeds*
1975-76	88.50	13.00	7.80	10.70	Nil	25.00	30.25	3.70	17.10	11.37	12.58	Nil
1976-77	68.90	13.10	7.30	10.70	Nil	25.75	30.320	5.37	14.80	11.58	12.30	Nill
1977-78	69.30	12.32	7.38	10.60	Nil	27.00	30.35	5.00	15.05	10.50	12.10	Nil
1978-79	70.00	12.00	8.25	9.75	Nil	27.10	28.82	6.10	10.41	11.30	12.70	3.57
1979-80	74.35	10.15	6.50	9.00	Nil	28.30	27.35	7.53	10.01	10.51	12.30	4.00
1980-81	75.10	10.00	6.51	8.39	Nil	31.00	23.31	9.20	11.54	10.10	10.50	4.35
1981-82	78.53	8.71	5.66	7.10	Nil	32.30	24.53	10.35	11.62	9.90	11.30	Nil
1982-83	81.00	7.00	5.70	6.30	Nil	35.77	20.30	12.30	10.36	9.75	11.52	Nil
1983-84	82.78	6.75	4.39	6.08	Nil	35.00	18.00	15.10	7.51	9.50	11.39	3.50
1984-85	83.25	5.50	5.98	5.27	Nil	37.40	12.15	17.59	8.25	10.35	12.11	2.15
1985-86	84.52	5.32	5.16	5.00	Nil	40.00	8.50	19.25	7.60	9.80	12.32	2.78
1986-87	87.32	5.00	4.39	3.29	Nil	41.60	8.30	20.23	9.59	8.78	11.58	Nil
1987-88	89.70	3.72	2.19	4.39	Nil	41.50	5.32	22.17	10.54	9.00	10.10	1.37
1988-89	92.98	0.69	0.65	5.66	Nil	43.01.	3.87	23.49	9.23	9.94	10.27	0.29

Source : Village Revenue Records, Rohankhedal

Note : NSA - Net Sown Area, ANC - Area Not Available for Cultivation, FL - Follow Land, CW- Cultivable Land.

Spatial and Temporal Analysis Agricultural Landuse

Rohankheda has 532.10 hectares under cultivation, accounting for 92.98 per cent to total geographical area. The village is noticed 1.56 per cent irrigated land to net snow area while 4.18 per cent to net sown area is recorded double cropped area. The agricultural landuse in Rohankheda mainly depends on rainfall amounts and availability of resources at local level. Generally, like other villages, this village experiences two crop seasons, *namely*, kharif and rabi. Among these kharif has an immense importance due to larger areal extent is brought under cultivation in this season. The crops such as cotton, hybrid jowar mug, not, sesamum, tur, rice are successfully grown. Among these crops, cotton ranks first in areal extent occupying 43.01 per cent to net sown area, sown on clay and clay loan soil in all parts in Rohankheda. Improved varieties of cotton are sown, and therefore have high yield per hectare. Hybrid jowar is grown on 23.49 per cent to net sown area in both soil types in west, east and southeast in major three patches. Besides there three patches the cultivation of hybrid jowar appears in central part along the kachha road on clay loan soils (Exhibit 9.19). Others crops have less than 6 per cent to net sown area in Rohankheda in kharif season. Tur accounts 3.87 per cent to net sown area while, sesamum is of 6.52 per cent to net sown area. Tur is mixed and sown with cotton on extensive cotton fields. Sesamum is sown in three patches located in west, central and southwest within the village area. This crop is confined to clay loam. Rabi season season starts in October when wheat, gram and linseed are grown in the village. Wheat is major rabi crop, accounting for 8.48 per cent to net sown area and is grown on both clay and clay loam soil types in three isolated patches. Linseed and gram and occupy 3.75 per cent and 2.27 per cent to net sown area respectively.

Exhibit 9.21 and Table 9.36 reveal the temporal variations in agricultural landuse in Rohankheda from 1975-76 to 1988-89. It was observed that village has constant impetus on growing hectarage under crops in kharif season than that of rabi. Among the various crops, cotton, hybrid jowar show upward trend while, local jowar, other cereals and oilseeds have downward trend (Exhibit 9.21 and Table 9.36). The highest increase has found in the hectarage of hybrid jowar (19.79 per cent to net sown area) followed by cotton (17.99 per cent to net sown area) in Rohankheda. These two varieties of crops are of immense significance and they produce high yield per hectare. The remarkable reduction has observed in local jowar. Its reduction is of 26.38 per cent to net sown area. This large reduction has been due to introduction of hybrid jowar, high yielding variety in Rohankheda on 23.49 per cent to net sown area in 1988-89 and therefore, local jowar is replaced and is sown on 3.87 per cent in 1988-89. Oilseeds, other cereals pulses, and other miscellaneous crops have shown declining trend in Rohankheda (Exhibit 9.21). Among these crops, the largest reduction in areal extent is found in other cereals, accounting for 7.87 per cent to net sown area. It had 17.10 per cent to net sown area in 1975-75, while it declined to 9.23 per cent to net sown area in 1988-89. Other crops *namely,* pulses found decline of 3.43 per cent to net sown area followed by other miscellaneous crops, (2.28 per cent) in Rohankheda.

Table 9.37
Agricultural Landuse Pattern (1988-1989)

Sr. No.	*Name of crops*	*Percent to net sown area*	*Sr. No.*	*Name of crops*	*Percent to net sown area*
1.	Cotton	43.01	8.	Mot	0.58
2.	Hybrid jowar	23.49	9.	Chilli	0.10
3.	Local jowar	3.87	10.	Vegetables	0.19
4.	Tur	6.22	11.	Wheat	8.48
5.	Sesamum	6.52	12.	Linseed	3.75
6.	Mug	0.77	13.	Gram	2.27
7.	Rice	0.75			

Source : Village Revenue Records, Rohankheda.

Landuse and Population

86.98 per cent working force to total workers is engaged in agriculture. It indicates the pressure of population on land. Table 9.38 presents the share of per capita in Rohankheda. Per capita land to total area is of 0.73 hectare which reduced to 0.68 hectare per head of population under land available for cultivation. Furthermore, it was observed that per capita land is high than that of rabi crops. Area under kharif crops has 0.58 hectare and 0.09 hactare per head population in area under rabi crops. It explains the dominance of kharif crops in Rohankheda and these kharif crops are of immense significance to this village as major crops are grown in this season.

Table 9.38
Landuse and Population

(Population 780)

Sr. No.	*Landuse categories*	*Area in hectares*	*Land per head of population*
1.	Total village area	573.36	0.73
2.	Land available for cultivation	532.10	0.68
3.	Total kharif crop area	454.86	0.58
4.	Total rabi crop area	77.24	0.09
5.	Gross cropped area	547.06	0.70
6.	Double cropped area	23.96	0.03

Source : Author

Double cropped area accounts for 0.03 hectare land per head of population whereas it accounted 0.7 per cent to total gross cropped area in the village. Table 9.39 reveals the demographic characteristics. Population figure from this table shows steady increase during the span of fourteen years. It has 580 persons in 1977 while it recorded 780 persons in 1989 with density of 1.01 persons and 1.36 persons per hectare respectively. The total increase of density of population is 0.35 hectare persons in study period. Literacy shows highest percentage than that of other sample villages (53.20 per cent to total population) indicating positive change in increasing awareness regarding agricultural practice. Table 9.40 presents the breakup of occupational structure in Rohankheda. It is found that village has dominance of agricultural labourers from 1971 to 1989. Its number in 1971 was 66.82 per cent and 64.38 per cent in 1989. Cultivators are reduced by 2.50 per cent during the span of fourteen years. Among other categories, workers accounted little number. Marginal workers accounting for 3.88 and workers other than farm workers is of 9.13 in 1989.

Table 9.39
Demographical Characteristics

Sr. No.	*Categories*	*Years*		
		1971	*1981*	*1989*
1.	Total population	580	662	780
2.	Density of population	1.01	1.15	1.36
3.	Literacy percentage	36.03	51.20	53.20
4.	Per capita land	0.96	0.86	0.73
5.	Percentage of farm workers to total workers	93.92	94.78	96.89

Source : Wardha District Census Handbook, 1971, 1981 and Village Panchayat Office, Rohankheda.

Agricultural Operation and Yields

Table 9.41 reveals the agricultural operation and yields of five crops in Rohankheda. Among these five crops, four crops are sown in kharif and rest is in rabi. Kharif crops are sown generally in June on commencing southwest monsoon. The land for cotton, tur, hybrid, jowar, and sesamum are prepared by one ploughing and two harrowing in May. Cotton and tur in Rohankheda are sown by dibbling method while hybrid jowar and sesamum are cultivated by drilling method. Cotton is sown on extensive hectarage followed by hybrid jowar. This rainfied crop takes 4 and 1/2 months to mature itself producing 250 to 300 Kgs. yields per hectare (Table 9.41). Hybrid jowar is sown in June by drilling mixing with mug. It takes three months for maturity. Sesamum is grown on clay loam in July after sowing of hybrid jowar and cotton. It yields 125 kgs per hectare in Rohankheda. Wheat is raised in October-November on moisture retaining soils on clay and clay loam (Table 9.41) by drilling method.

Socio-Economic Facilities and Farmers

Majority of people in village is economically poor. Nine farmers were observed having more than 10 hectares and they raised improved varieties of cotton and hybrid jowar extensively while rest farmers sown these varieties on limited areal extent in fields. Poor farmers can not afford improved implements and improved seeds due to lack of capital to them. It is observed that samll land holding farmers borrow or hire agricultural implements from others. It unfolds during the interview that capital is the great handicap in the development of agriculture. Moreover, these small land holders are interested to make substance development in proper utilization of land. Absence of market, high school, bus services, medicals are somewhat abstracle in the way of improving agriculture.

Resume

This village is representative of developed agricultural region involving 86.98 per cent in agricultural practice. It is noteworthy that farmers of this village are aware of new development in the field of agriculture. Cotton and hybrid jowar are common crops grown extensively (43.01 per cent to net sown area and 23.49 per cent to net sown area) on clay and clay soils. Larger size land holders are greatly advanced in improving agricultural land by producing high yield of cotton and hybrid jowar. These farmers raised rice and vegetables by well irrigation. While other farmers who owned 2 to 5 hectares land cultivated local jowar, mug, mot, groundnut, gram, sesamum, besides, cotton and hybrid jowar. By increasing availability of capital and providing latest information to farmers on agriculture. Rohankheda may serve proper landuse for increasing yield per hectare.

Table 9.40
Occupational Structure

Sr. No.	*Categories*	*Population*			*Percent of total population*			*Percent of total workers*		
		Years								
		1971	*1981*	*1989*	*1971*	*1981*	*1989*	*1971*	*1981*	*1989*
1.	Cultivators	62	77	99	10.68	11.63	12.60	25.10.	21.44	22.60
2.	Agricultural labourers	170	250	282	29.31	37.76	36.15	68.82	69.63	64.38
3.	Marginal worker	4	14	17	0.68	2.11	2.17	1.61	3.89	3.88
4.	Workers other than Agricultural labourers	11	18	40	1.89	2.71	1.74	4.45	5.01	9.13
5.	Total workers	247	359	438	42.58	54.22	—	100.0	100.0	100.0
6.	Non-workers	333	303	342	57.41	45.77	43.34	-	-	-
7.	Total population	580	662	780	100.0	100.0	100.0	-	-	-

Source : Wardha District Census Handbook for 1971, 1981 and Village Panchayat Office, Rohankheda.

Table 9.41
Agriculture and Yields

Landuse	*Crops*				
Operations	*Cotton*	*Hybrid Jowar*	*Tur*	*Local Jowar*	*Groundnut*
Botanical Name	Gossypium spp.	Sorghum Bicolor	Triticum vulgare	Sesamum indicum	Cajanus indicus
Soil	Clay/ Clay loam	Clay/ clay loam	Clay/ clay loam	Clay loam	Clay loan/ clay
Sowing time	June/July	June	Oct-Nov.	July	June/July
Spacing	3' to 4'	7" to 9"	5" to 7"	9" to 10'	3" to 4"
Seed rate per acre in kg.	1 Kg	1Kg	10 to 15 Kg.	2 Kg.	2½ Kg.
Method of sowing	Dibbling	Drilling	Drilling	Drilling	Dibbling
Manures of Fertilizers (NPK) in Kg. per acre	30-15-0	10-5-0	Nil	Nil	Nil
Irrigation	Rainfed (Kharif)	Rainfed (Kharif)	Rainfed (Kharif)	Rainfed (Kharif)	Rainfed (Kharif)
Harvesfing time	4 to 4½ months	3 months	5 months	2½ to 8 months	4 months
Yield per acre in kg.	250 to 300 Kg.	200 kg.	250 kg.	125 kg.	150 kg.

Source : Field work

6. SARUR

Introduction

The village Sarur is located in central part with an areal extent of hectares (334.37) in the Yashoda basin (Exhibit 9.1). It is 18 kms away from Wardha in South and 8 kms from Kangaon on Wardha-Kangaon metalled road. Moreover, Sarur links with surrounding village by kachha roads (Exhibit 9.22). It is noticed from Exhibit 9.22 that this village is confined by Borgaon and Mamdapur in south, Borgaon in eat, Takali in west, and Jaitapur and Sekapur in north. It is situated on the right bank of river Yashoda and is located on 20°.25' North latitude and 78°.37' East longitude. The general slope is towards east which is easily noticed from the water flow of Yashoda (Exhibit 9.22) The gaothan is located on right bank of river Yashoda. The village had 527 population in 1981 and 662 in 1989 with density of 1.57 persons and 1.97 persons per hectare respectively. The observations and interviews unfold that the Teli and Kunbi caste are engaged in agriculture and scheduled caste and tribes are agricultural labourers.

Climate and Soils

Soil types and slight variations in amount of rainfall are continuously influencing agricultural landuse. Intensive heat in summers and farily cold in winters are common in Sarur. The village receives rainfall from southwest monsoon. The rainfall in kharif season is of great importance in growing varieties of crops. Three types of soils are observed in Sarur. These are namely, sandy clay loams, clay loam and clay (Exhibit 9.22). The analysis of soil types is given in Table 9.42.

The largest coverage is found of clay soils spread over an extensive area in north part. This soil contains 48 per cent clay and 13 per cent silt and produces than that of other soils types in Sarur. Cotton and hybrid jowar are successfully grown on this soil. It covers 38 per cent to total village area. Clay loam appears in central west and south west occupying 35 per cent of total village a area. It is fertile soil containing 40 per cent silt and 35.50 per cent clay. It is observed that this soil type has highest pH value of 8. Sandy clay loam spreads over 22 per cent to total village area in east of Nalla in south-west. This is of inferior soil type than that of earlier soil types having 34.85 per cent fine sand and 15.10 per cent coarse sand on which hardy crops like local jowar, mug, mot, sesamum are grown successfully.

Table 9.42
Soil Analysis of Sarur

Soil texture classes	*Soil Types*		
	Clay (%)	*Clay loam (%)*	*Sandy Clay loam (%)*
Coarse sand	11.90	8.10	15.10
Fine sand	22.08	12.87	34.85
Silt	13.75	40.00	11.25
Clay	48.00	35.50	33.00
Nitrogen	0.51	0.36	0.29
Phosphate	1.36	0.56	1.72
Potash	24.00	16.25	20.50
pH Value	7.9	8.0	7.8

Note : The above analysis was carried out in the Soil Testing Laboratory, Department of Agriculture, Pune 411005.

Spatial and Temporal Analysis of General Landuse

Table 9.43 exhibits general landuse types in Sarur for 1988-89. This village has 288.60 hectares land under cultivation (86.31 per cent to total geographical area). Except few patches along river (Exhibit 9.23). Crops are raised on extensive area despite soil types. The major concentration of cultivated land is observed in central and western parts where clay and clay loam soils are dominant and productive to yield high crops. Spot inquiries reveal that two farmers have tried to make intensive cultivation in fields located on river Yashoda.

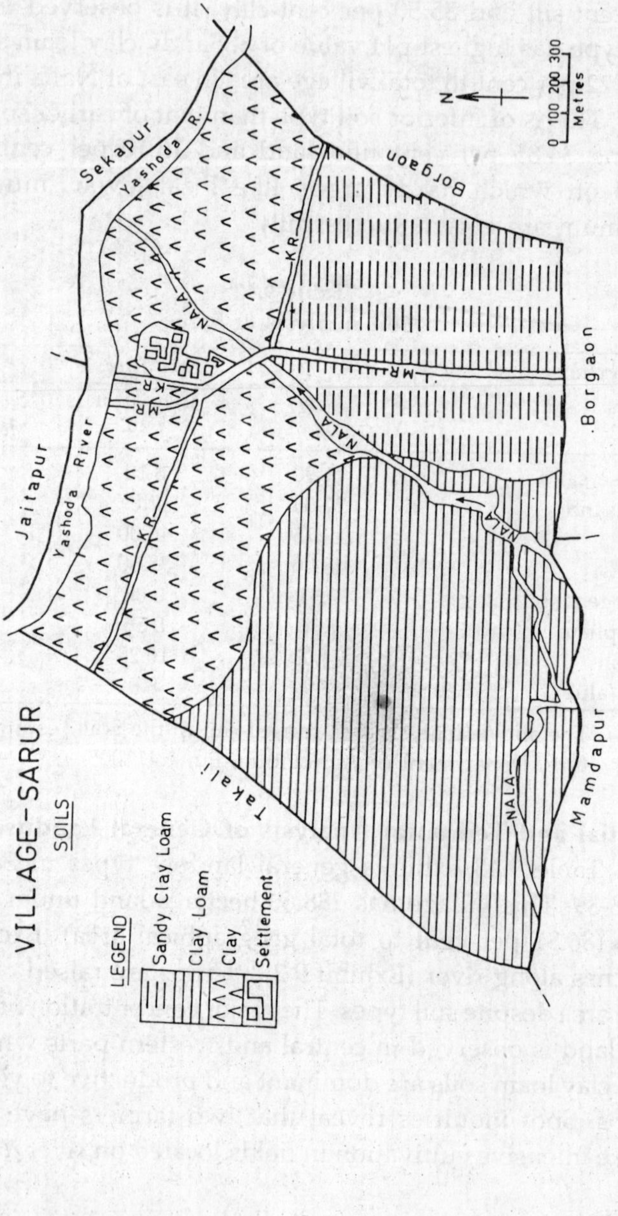
VILLAGE-SARUR
SOILS
LEGEND
Sandy Clay Loam
Clay Loam
Clay
Settlement
Jaitapur
Yashoda River
Sekapur
Yashoda R.
Borgaon
Borgaon
Mamdapur
Takali
NALA
KR
MR
N
0 100 200 300
Metres

EXHIBIT — 9.22

Table 9.43
General Landuse Patterns (1988-89)

Sr. No.	Landuse types	Area (hectares)	Percent of total area.
1.	Net Sown Area	288.60	86.31
2.	Cultivable Waste	21.30	3.37
3.	Fallow Land	10.70	3.20
4.	Land not available for cultivation	14.77	4.11
5.	Forest	Nil	Nil
6.	Double Cropped Area	9.39	3.25
7.	Total Cropped Area	297.99	89.56
8.	Irrigated Area	1.95	0.58
9.	Total Geographical Area	334.37	—

Source : Village Revenue Record, 1988-89.

Among other categories of landuse types, cultivable waste accounts highest land in the village in two patches concentrated in northwest part (Exhibit 9.23) due to dense shrubs and trees on rivers bank of Yashoda. It occupies 6.37 per cent to total geographical land not available for cultivation accounts for 4.11 per cent to total geographical area in northwest part on river bank where local hilly, rugged land in dominant. Fallow land is observed in clay soil tract accounting for 3.20 per cent to total geographical area. Forest as defined by forest department is almost absence except disperse trees and shrubs. Few farmers have observed to irrigate land occupying 0.58 per cent to net sown area.

Table 9.44 summarised the general landuse types from 1975-76 to 1988-89 in Sarur. The most remarkable fact that net sown area is steadily increased during the span of fourteen years while other landuse categories declined hectrage and show downward trend. The total increase of net sown area is 21.01 per cent to total geographical area from 1975-76 to 1988-89. It has 65.30 per cent to total geographical area under net sown area while it is increased to 86.31 per cent to total geographical area in 1988-89. This fact unfolds the awareness of farmers to bring more land under cultivation. Other landuse types have shown downward trend during the study period.

The highest decline is observed in land not available for cultivation by 11.21 per cent to total geographical area. Fallow land is declined by 5.40 per cent to total geographical area followed by cultivable waste (4.41 per cent to total geographical area) in Sarur.

Spatial and Temporal Analysis of Agricultural Landuse

Like other sample villages, two agricultural seasons are prevailed in Sarur. Among these, season kharif is of great importance because more than 80 per cent crops are sown and produced in this season. Cotton, hybrid jowar, tur, mug, mot, sesamum and local jowar are kharif crops and are sown in June on commencing southwest monsoon. Among kharif crops cotton stands as first accounting 51.76 per cent to net sown area. This crop is mainly sown on clay and clay loam soils and therefore has high yield per hectare than that of other crops in Sarur. Hybrid jowar occupies 27.75 per cent to net sown area and stands second in rank. Improved varieties of hybrid jowar are sown in the village. It is uncovered during field work and personal inquiry that local jowar was replaced by hybrid jowar due to having comparatively high yield per hectare and because of this reason, the hectarge under local jowar has remarkably reduced and it stands fifth in rank occupying 3.50 per cent to net sown area. Tur is mixed sown with cotton. It is sown after twelve rows of cotton. It accounts for 7.32 per cent to net sown area. Other crops grown in kharif have insignificant hectarge (Exhibit 9.23). Rice occupies 0.79 per cent to net sown area followed by sesamum (0.33 per cent) and mug (1.03 per cent) in the village. Rabi season commences in October when wheat is grown and linseed are sown. Among these crops, linseed is grown on 4.67 per cent to net sown are and stands first in rank in rabi season. Wheat and gram occupy little hectarge of 1.17 per cent and 0.46 per cent to net sown area respectively.

Exhibit 9.25 and Table 9.46 presents temporal variations in agricultural landuse pattern during the study period (1975-96 to 1988-89). It is noticed from Exhibit 9.25 and Table 9.46

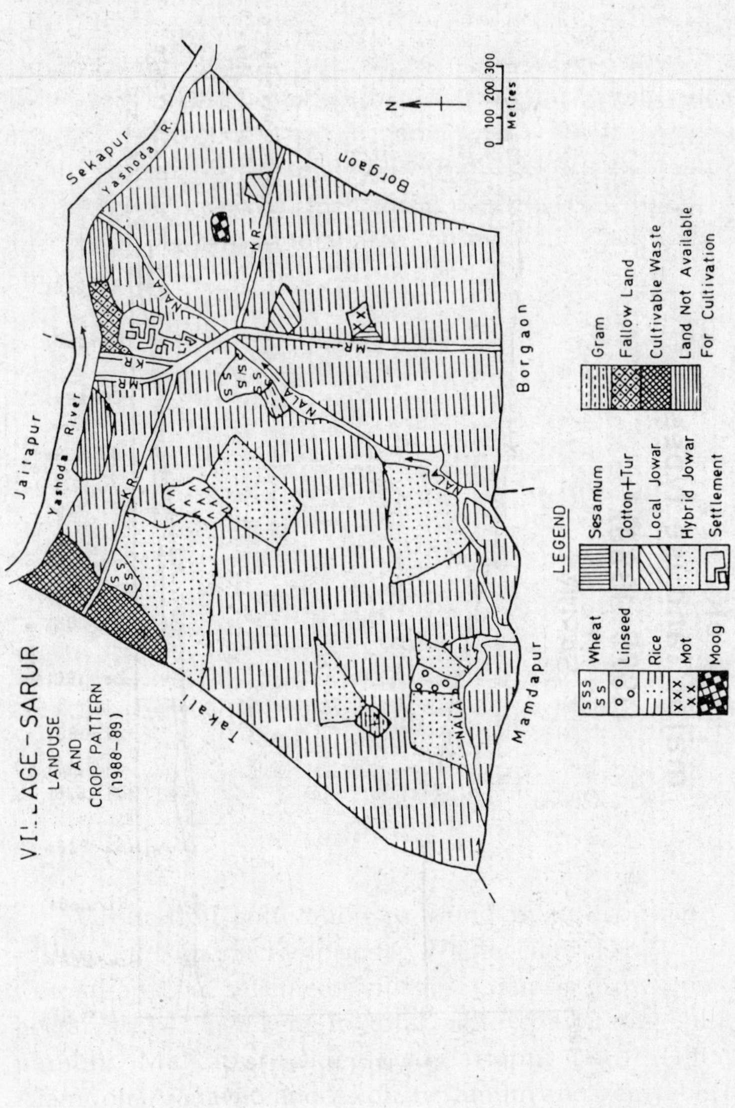
VILLAGE - SARUR
LANDUSE
AND
CROP PATTERN
(1988-89)
Sekapur
Yashoda R.
Jaitapur
Yashoda River
Borgaon
Borgaon
Mamdapur
Takali
NALA
KR
MR
N
0 100 200 300
Metres
LEGEND
Wheat
Linseed
Rice
Mot
Moog
Sesamum
Cotton+Tur
Local Jowar
Hybrid Jowar
Settlement
Gram
Fallow Land
Cultivable Waste
Land Not Available For Cultivation

EXHIBIT — 9.23

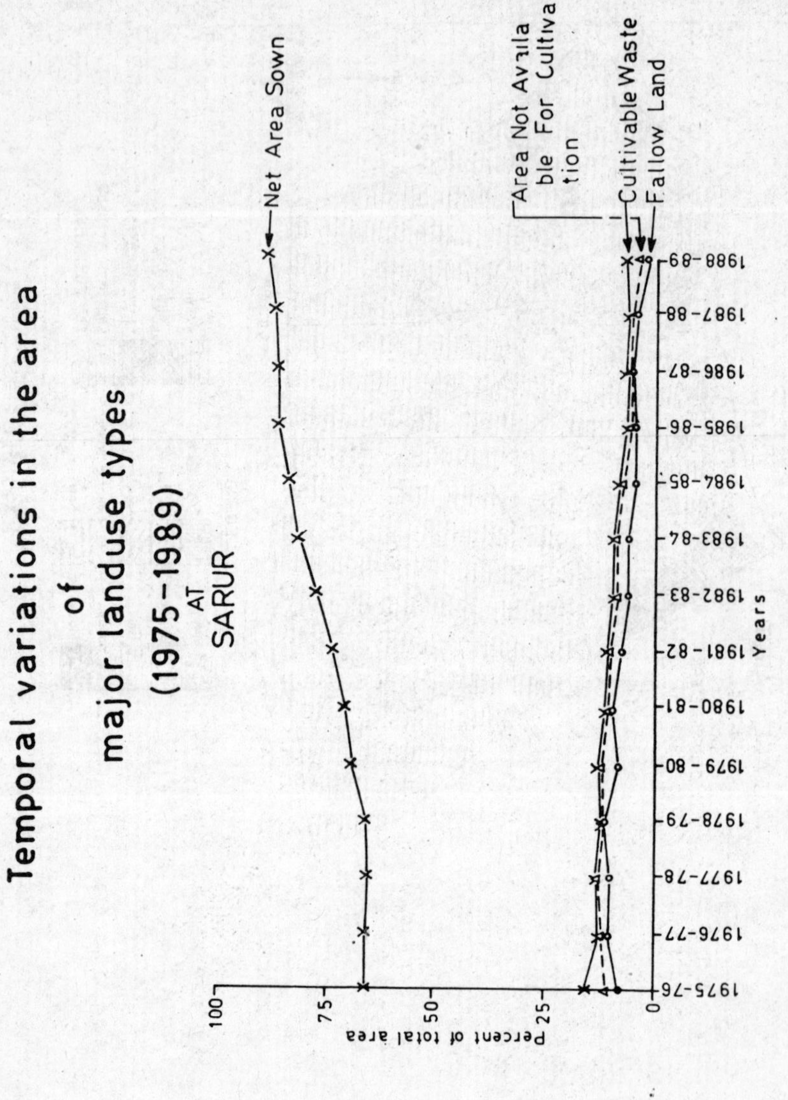
Temporal variations in the area
of
major landuse types
(1975-1989)
AT
SARUR
Net Area Sown
Area Not Availa ble For Cultiva tion
Cultivable Waste
Fallow Land
Percent of total area
100
75
50
25
0
1975-76
1976-77
1977-78
1978-79
1979-80
1980-81
1981-82
1982-83
1983-84
1984-85
1985-86
1986-87
1987-88
1988-89
Years

EXHIBIT — 9.24

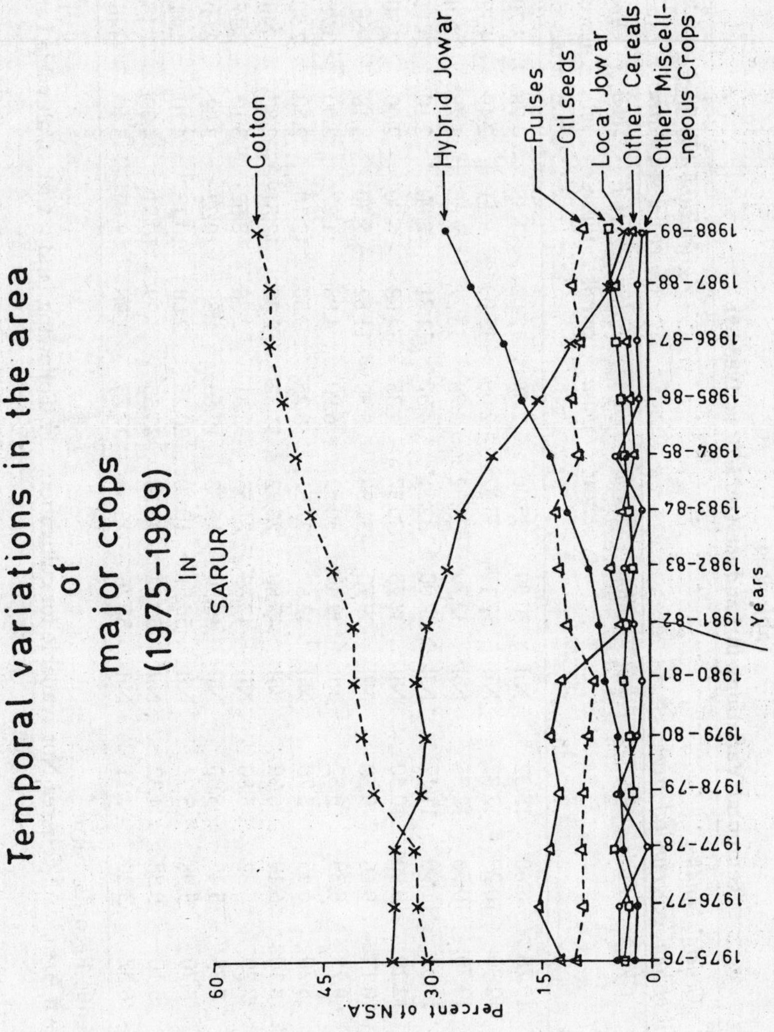

Temporal variations in the area
of
major crops
(1975-1989)
IN
SARUR
Cotton
Hybrid Jowar
Pulses
Oil seeds
Local Jowar
Other Cereals
Other Miscell-
-neous Crops
Percent of N.S.A.
60
45
30
15
0
Years
1975-76
1976-77
1977-78
1978-79
1979-80
1980-81
1981-82
1982-83
1983-84
1984-85
1985-86
1986-87
1987-88
1988-89

EXHIBIT — 9.25

Table 9.44
Temporal Variations in Landuse (1975-76 to 1988-89)

Years	*General Landuse (percent of total geographical area)*					*Agricultural Landuse (% of net sown area)*						
	NSA	*CW*	*FL*	*ANC*	*Forest*	*Cotton*	*Local Jowar*	*Hybrid Jowar*	*Other cereals*	*Total Pulses*	*Total oil*	*Other miscellaneous seeds*
1975-76	65.30	10.78	8.60	15.32	Nil	30.59	35.00	2.58	12.00	11.35	4.78	3.70
1976-77	65.11	12.30	10.27	12.32	Nil	32.11	35.17	2.00	15.00	9.11	2.49	4.12
1977-78	64.39	12.70	10.56	12.35	Nil	32.70	35.32	3.78	13.77	9.65	4.78	Nil
1978-79	65.52	11.32	11.90	11.17	Nil	37.00	31.11	4.87	11.81	8.90	2.10	5.11
1979-80	68.50	12.10	8.90	10.50	Nil	38.25	30.57	4.35	14.03	8.30	2.00	2.50
1980-81	70.73	10.11	9.66	9.50	Nil	39.70	31.38	5.80	11.82	7.30	4.18	Nil
1981-82	72.51	9.78	8.39	9.32	Nil	39.45	30.52	6.90	4.98	11.37	3.00	3.78
1982-83	76.15	8.70	6.45	8.70	Nil	42.67	27.51	8.25	3.50	12.10	2.50	2.90
1983-84	79.32	8.10	6.19	6.39	Nil	45.80	25.22	11.32	2.35	12.51	2.80	Nil
1984-85	81.50	7.39	5.32	5.80	Nil	47.33	21.55	13.31	2.38	8.62	3.10	3.17
1985-86	84.17	5.38	5.45	5.00	Nil	49.30	15.33	17.68	1.85	10.32	3.52	2.00
1986-87	84.00	5.70	5.98	5.32	Nil	50.51	10.30	20.25	2.60	10.13	4.10	2.11
1987-88	84.70	5.10	5.88	4.32	Nil	50.32	5.18	23.55	4.99	10.51	4.30	1.15
1988-89	86.31	6.37	3.20	4.11	Nil	51.76	3.53	27.75	1.96	9.60	5.40	Nil

Source : Village Revenue Records, Sarur

Note : NSA Net Sown Area, ANC -Area Not available for cultivation, Fl -Following land, CM - Cultivable Land.

that two crops *namely,* cotton and hybrid jowar have shown upward trend while other crops have slightly declined during the span of fourteen years. The total decline of cotton during period is 21.17 per cent to net sown area. It had 30.59 per cent to net sown area in 1975-76 which increased to 51.76 per cent to net sown area in 1988-89. Hybrid jowar is found increased by 23.17 per cent to net sown area. It was sown on 2.58 per cent to net sown area on experimental basis which is yielded high production per hectare and therefore after 1975-76 the hectarage of hybrid jowar is steady increased and it is sown on 27.75 per cent to net sown area in 1988-89. Thus, both cotton and hybrid jowar due to high yield per hectare have readily adopted by villages and therefore show upward trend during the study period. All other crops such as local jowar, other cereals, oil seeds, and pulses are declined due constant increase in hectarage under cotton and hybrid jowar. The highest decline is observed in local jowar 31.47 per cent to net sown area followed by oilseeds (10.04 per cent), other miscellaneous crops (2.55 per cent) and pulses (1.65 per cent) in the village.

Table 9.45
Agricultural Landuse Pattern (1988-1989)

Sr. No.	*Crops*	*% of NSA*	*Sr. No.*	*Crops*	*% of NSA*
1.	Cotton	51.78	7.	Rice	0.79
2.	Hybrid jowar	27.75	8.	Mut	1.07
3.	Local jowar	3.53	9.	Chilli	0.00
4.	Tur	7.32	10.	Wheat	1.17
5.	Sesamum	0.38	11.	Linseed	4.67
6.	Mug	1.03	12.	Gram	0.48

Source : Village Revenue Records, 1988-89.

Landuse and Population

Table 9.46 exhibits per capita share of land to various categories of landuse in Sarur. The per capita land is 0.50 hectare in the village and per capita land for cultivated land is 0.43 hectare. This land further reduced into kharif and rabi area. The area under kharif crops show 0.40 hectare per head of population while 0.02 hectare is observed in the area under rabi (Table 9.46). It unfolds the dominance and population concentration on kharif area than that of rabi. The demographic characteristic are show in Table 9.47. It is clear from this table that population increases during the period of study. It was noticed 482 persons in 1971 which is attained to 662 persons in 1988-89. The density is also increased by 0.53 persons per hectare. It had 1.44 persons in 1972 and in the 1989 it is found 1.97 persons per hectare. Literacy percentage shows upward trend during the period of fourteen years.

Table 9.46
Landuse and Population

(Population 662)

Sr. No.	*Landuse categories*	*Area in hectares*	*Land per head of population*
1.	Total village area	334.37	0.50
2.	Land available for cultivation	288.60	0.43
3.	Total kharif crop area	270.31	0.40
4.	Total rabi crop area	18.29	0.02
5.	Gross cropped area	297.95	0.45
6.	Double cropped area	9.39	0.01

Source : Author

Table 9.47
Demographical Characteristics

Sr. No.	*Categories*	*Years* 1971	1981	1989
1.	Total population	482	527	662
2.	Density of population	1.44	1.57	1.97
3.	Literacy percentage	25.51	48.19	49.09
4.	Per capita land	0.69	0.63	0.50
5.	Percentage of farm workers to total workers	82.52	84.45	93.15

Source : Wardha District Census Handbook, 1971, 1981 and Village Panchayat Office Sarur.

It increased by 23.50 per cent. It has increased nearly double to base year *i.e.* 1971 (Table 9.48). Both cultivators and agricultural labourers increase in percentage during study period. It uncovers the growing significance of agriculture among farmers. It is noticed that 35.95 per cent cultivators and 57.19 per cent agricultural labourers have been registered in 1988-89. Marginal workers and workers engaged other than farm workers and workers accounts fro 6.16 per cent and 0.68 per cent to total workers in Sarur.

Agricultural Operation and Yields

Two agricultural seasons prevail in Sarur, *namely,* kharif and rabi. Kharif fields are ploughed in May by iron plough and wooden hurrows. Much concentration in paid in preparing cotton and hybrid jowar fields as they produce high yield per hectare and are grown on extensive area. Cotton is sown in June while hybrid jowar in sown after completion of cotton sown. Dibbling method is common in cotton sowing while hyrbid jowar is sown by drilling method. Cotton, Hybrid jowar, tur, and local jowar are rainfed crops. Cotton takes three to four moths for maturity while hybrid jowar is matured in three months. Comparatively cotton produces high yield per hectare than hybrid jowar in Sarur. Other crops *namely* tur and local jowar have little significance hence are sown on variety of soils and therefore have low yield per hectare. Linseed is sown on moisture retaining soils in rabi season by drilling method in Sarur.

Socio-Economic Facilities and Farmers

It is found that Teli and Kunbi are engaged as cultivators while scheduled castes and tribes are involved as agricultural labourers. Two farmers are observed having more than ten hectares land and are well developed. These farmers produce cotton and hybrid jowar on extensive area. This could be possible to them because of large size of land holding. Other people having small size of land holding can not spend

Table 9.48
Occupational Structure of Sarur

Sr. No.	Categories	Population			Percent of total population			Percent of total workers		
		Years								
		1971	1981	1989	1971	1981	1989	1971	1981	1989
1.	Cultivators	62	62	105	12.86	11.76	15.86	30.09	32.12	35.95
2.	Agricultural labourers	116	121	167	24.48	22.96	25.22	57.28	62.69	57.19
3.	Marginal worker	-	0	18	-	-	2.71	-	-	6.16
4.	Workers other than agricultural	26	10	2	5.39	1.88	0.30	12.62	5.18	0.68
5.	Total workers	206	113	292	42.73	36.62	44.10	100.0	100.0	100.0
6.	Non-workers	276	334	370	57.26	63.37	55.89	-	-	-
7.	Total	482	527	662	100.0	100.0	100.0	-	-	-

Source : Wardha District Census Handbook for 1971, 1981 and Village Panchayat Office, Sarur.

Table 9.49

Agricultural Operation and yields for major crops of Sarur

Landuse Operations	*Crops*				
	Cotton	*Hybrid Jowar*	*Tur*	*Local Jowar*	*Wheat*
Botanical Name	Gossypium Spp.	Sorghum bicolor	Cajanus indicus	Sorghum vulgare	Triticum vulgare
	loam/ sandy clay	loam/ sandy clay	loam/ sandy clay	loam/ sandy clay	clay
Sowing time	June/July	June	June/July	June	June /July
Spacing	4 to 4½'	7 to 9"	4 to 4½"	5 to 7"	7 to 10"
Seed rate per acre in Kg.	1Kg.	1Kg	2½ Kg.	3 Kg	2½ Kg
Method of sowing	Dibbling	Drilling	Dibbling	Drilling	Drilling
Manures of Fertilizers (NPK) in Kg. per acre	35-5-0	9-5-0	Nil	Nil	Nil
Irrigation	Rainfed (Kharif)	Rainfed (Kharif)	Rainfed (Kharif)	Rainfed (Kharif)	Rainfed (Kharif)
Harvesting time	3 to 4 months	3 months	4 months	3½ months	3 months
Yield per acre in Kg.	200-250 Kg.	150 Kg	100 Kg	150-175	200 Kg

Source : Village Revenue Records, Sarur.

money on purchasing improved seeds and agricultural implements. Small size landholders borrow or hire these from neighbourers. Fragmented land in Sarur is also hindrance in proper and efficient landuse. Fragmentation could be attributed to division and sub-division of families in the village.

Resume

Sarur is representative of developed agricultural region, involving 93.15 per cent working force in agricultural on fertile soil of clay and clay. Hybrid jowar and cotton are successfully grown. Cotton accounts fro 51.75 per cent to net sown area while hybrid jowar is grown on 27.75 per cent to net sown area. Other crops, tur found wheat are sown in all soil types and production of these crops mainly depend on amount of rainfall and soil types. Tur accounts 7.32 per cent to net sown area. It is noticed that farmers are aware of development of agriculture. Due to deep black soils digging of well is expensive and risky. But proper planning of application of fertilizers and pesticides, can serve to improve village and field in Sarur.

10

Summary and Observations

INTRODUCTION

In recent years, efforts have been made to analyse and find out the nature of soils and their characteristics for suitable and proper growth of crops and their productions. Such a study is heavily dependent on scientific and comprehensive study of landuse at macro and micro levels. In order to comprehend the landuse patterns, surveys have been carried out in Britain, United States of America, China, Japan, Italy, France, Poland, Pakistan, Bangladesh and U.S.S.R. at regional level. Such surveys carried out at regional level do not fully justify to uncover landuse characteristics at village level. The present study is therefore significant in this respect. This study concentrates at the micro level on the village as the selected unit.

The present investigation has been carried out to assess, analyse, describe and interpret the pattern of agricultural landuse of the Yashoda basin (Wardha District) with a view to investigate the influence of certain physical and socio-economic variables on landuse patterns and to provide an indepth study for adjudging scientific, proper and efficient landuse to meet the demand of food for the increasing population in the basin and to demarcate agricultural regions for planning.

Summary and Major Findings

The Yashoda basin a part of Wardha District, Maharashtra extending from 20°.20' to 20°.45' North and 78°.45' to 79°.00 East. The basin comprises 155 villages of Dewali, Arvi, Hinganghat and Wardha talukas of Wardha district. It has the length of about 24 kms from north to south and the width of 14.5 kms from west to east and covers an area of 860.88 Sq. Kms. encompassed 13.64 per cent of Wardha District and 0.27 per cent area in Maharashtra, with total population of 11436 persons, according to the census of 1981 and density of 92 persons per square kilometre. Physiographically, the entire basin presents remarkable homogeneity in character with moderate slope. It can be divided into three physiographic units, *namely,* the Tigaon Hill, the Bhadadi and Punsoda doab, and the lower Yashoda basin (Exhibit 3.3). The Tigaon Hill region is relatively rocky and rugged with interior soils than that of the other two physiographic units. The Bhadadi and Punsoda doab as well as the lower Yashoda basin have plain surfaces with medium to deep black soils (Exhibit 3.7), yielding high production per hectare. The rivers Yashoda, Bhadadi and Punsoda constitute the drainage system in the study region. All three rivers are rainfed and seasonal in character. Transported fine grains and clays by the Yashoda are deposited in the downstream regions and form alluvial aprone where varied crops are successfully grown. The tributaries of Yashoda namely, Bhadadi and the Punsoda rivers are equally significant agriculturally. The typical monsoon climate with small variation in rainfall and range of temperature prevail. The region experiences intense heat in summer and fairly cold winter. The summers are dry which cannot promote the growth of vegetation. The mean monthly temperature is 20°.9C in January and 30°.8C in May. The average annual rainfall has 700 mms with slight variations in its amounts over the basin. The success of agricultural operations growth of crops, and their production depends on regular and sufficient rainfall during the rainy period. Rainfall in the basin between June and October with an average monthly rainfall of 900 mms. The rainfall amount diminishes after November with occasional stormy spells. The soil types are key parameter in the basin. Three types of soils, *namely*, coarse shallow, medium black and deep black soils with varied texture, colour, structure and fertility (Table 3.5).

The observations, analysis and findings discussed in the preceding chapters are presented below:

1. The Yashoda basin is predominantly agricultural in its occupational structure with 87.57 per cent working force to total workers engaged in agriculture. The types of soils play a vital role in agricultural landuse patterns in the basin. Even slight variations in rainfall amount affects agricultural landuse in the area under review.

2. The socio-economic variables have influenced the spatial distribution of agricultural landuse patterns in the Yashoda basin. The spatial distribution of population reveals the significant variations within the basin. The major concentration of population is associated with medium to deep black soil zones in central, eastern and southern parts of the region. The density variations in the basin range from 11.90 persons per square kilometer to 192 persons per square kilometre. The eastern and central parts have dense population due to fertile soils. With growing population in the basin, the pressure on land has been increasing and therefore per capita land has significantly decreased from 1971 to 1981. It had 0.58 hectare per capita land in 1971 and it decreased 0.37 hectare per capita land in 1981. The relative importance of agricultural activity has been pointed out by working force (87.51 per cent to total workers) in the basin, comprising 33 per cent of cultivators and 40.70 per cent agricultural labourers. The size of land holding has been adversely affected by continuous disintegration of land holdings due to division and sub-division of family. The average size of holding is 3.5 hectares. Traditional implements are still in use. The total irrigated area is 0.58 per cent to net sown in the Yashoda basin. Well irrigation is the only source of water supply to crops significantly in the central and eastern parts. The highest area under irrigation is recorded at Dewali (29.3 per cent to net sown area). The accessibility by means of transport, within the basin is easy, a dense road network, covering major district road, other district road and kachcha roads occur in the area. The western part has a comparatively sparse road network due to its relatively underdevelopment and low fertility of soils. Eighteen market centres serve to agglomerates and distribute the agricultural products and people within and outside the basin. Bus services in remote villages are seasonal.

3. This land utilization study unfolds different uses in the Yashoda basin. The net sown area manifests spatial variations within the basin, with concentration in the southern and eastern parts in fertile tract having high yield per hectare. Very few patches scattered in areal extents on coarse shallow soils have low concentration of net sown area with resultant low yield of crops per hectare. Chanaki has recorded the maximum net sown area (91.5 per cent to total geographical area) and the minimum is observed at Kachanpur (41.33 per cent to total geographical area) in the northwest part in the basin. It is noticed that the central tract of the basin is 70 to 80 per cent of net sown area in each village. The temporal component of land utilization demonstrates changes during the decade covering the years 1975-76 to 1985-86 in the Yashoda basin. The net sown area in the basin is 79.58 per cent to the total geographical area. It is relatively more in percentage than that of Wardha district (66.92 per cent) to total geographical area and Maharashtra (59.49 per cent) to total geographical area. The total increase in net sown area during the study period is 30.19 per cent to the total geographical area due to the increasing awareness among the farmers of the region to the improved varieties of seeds, fertilizers and pesticides for getting higher yields. Other categories of landuse have been found to decline steadily (Exhibit 5.1). The highest land under cultivable waste has declined to 7.07 per cent to total geographical area followed by fallow land (4.49 per cent to total geographical area) and 2.31 per cent to total geographical area) in the Yashoda basin.

4. The cultivation of crops in the Yashoda basin is confined to kharif and rabi seasons. It is noticed that kharif season is important in raising varieties of crops, with larger in areal extent in the basin. It is observed that more than 70 per cent to the net sown area is under kharif crops. The largest areal extent has been recorded under cotton (38.50 per cent to net sown area) with heavy concentration in the central and southern parts in the basin on medium to deep black soils stands first in rank followed by hybrid jowar on 16.88 per cent to net sown area in the basin. Hybrid jowar is widely distributed within the basin. Local jowar is raised on 11.74 per cent to net sown area followed by rice (3.43 per cent to net sown area), tur (2.89 per cent to net sown area), and mot (2.26 per cent to net sown area). Wheat is the leading rabi crop sown in the basin covering 12.77 per cent to net sown area. Other crops are insignificant in areal extent and hence have little impact on the overall agricultural landuse in the basin. Mot and mug are generally sown on coarse shallow soils of low fertility in the area under review. The temporal study reveals the present trend of cropping pattern. The study demonstrates that two crops *namely*, cotton and hybrid jowar have witnessed a remarkable upward trend as compared to other crops raised in the Basin. (Exhibit 6.1). The total increase in the percentage area under cotton during the study period is 17.11 per cent to net sown area than the base year that is, 1975-76. This may be attributed to the use of improved high yielding varieties of seeds for sowing. The increase in hybrid jowar during the period, is 20 per cent to net sown area. Other crops, *namely*, local jowar, mug, mot, linseed, sesamum have recorded a steady decline in the basin since 1980. These crops can hardly compete with hybrid jowar and cotton. Local jowar has declined by 20.00 per cent followed by other cereals (10.04 per cent) and oilseeds (2.89 per cent) to net sown area during the time span covering in this study.

5. For the exposition of agricultural landuse pattern of the Yashoda basin, the village has been chosen as a unit for study to find out the crop combination and diversification. The ranking of crops denotes the relative strength of individual crops. Cotton ranks as the first ranking crop in one hundred and fourtysix villages (94.19 per cent) to total villages) and consequently occupies an outstandingly predominant position in 26387.95 hectares. Hybrid jowar stands second in rank in one hundred and twelve villages (72.25 per cent to total villages). Other crops, *namely*, local jowar, gram, wheat and rice have low percentage than that of cotton and hybrid jowar. Five crop combinations have been identified in the basin. Among five crop combinations three crop combinations are relatively dominant in areal extent in eighty villages (51.61 per cent to total villages). The sole village at Mankapur is noted for its monoculture, located in the northwest part with the prevalence of coarse shallow soil where local jowar is raised in larger areal extent, while five crop combinations obtained at Renukapur with combined local jowar, wheat, mug, gram, and sesamum in this combination. Further it is noticed that three and four crop combinations have wide distribution where cotton, hybrid jowar and local jowar are common constituent crops in these combinations. The existing diversifications of crops serve to unfold the competition among them. The resultant crop diversifications registered moderate increase in magnitude, covering eighty village (51.61 per cent to total villages) on 51239 hectares, followed by low degree of crop diversification in seventytwo villages (46.45 per cent to total villages) on 34519 hectares. It has been noticed that the riverine tracts of the Bhadadi and the Punsoda where cotton, hybrid jowar, wheat and tur entering into the moderate crop diversification. The southern and western parts on medium to deep black soils are noted for low crop diversification where cotton, hybrid jowar, wheat, sesamum and linseed comprise this class. Very high crop diversification is noticed at Renukapur, whereas very low diversification of crop at Nagazari and Bidnagazari in northwest part on coarse shallow soils, with low yields per hectare in the basin.

6. In order to evaluate the relationship between agricultural landuse variables and environmental variables in the Yashoda basin by applying contitative techniques *namely*, correlation coefficient, multiple regression, and factor analysis techniques have been used in the present investigation and finally agricultural regions have been identified. The correlation coefficient establishes relationship between chosen variables. The strong positive correlation was found between net sown area and hybrid jowar (0.600), local jowar and mug (0.755), local jowar and slope (0.645), irrigation and agricultural density (0.566), population density and agricultural density (0.945). This fact explains that areal extent of above crops are increasing with increasing area under net sown area and therefore the variables show positive correlation. Negative correlation established between net sown area and local jowar (-0.560), cotton and local jowar (-0.870), cotton and mug (-0.637), cotton and other pulses (-0.482), cotton and slope (-0.691), local jowar and wheat (-0.551), local jowar and tur (-0.478), wheat and tur (0.550). These variables are not experiencing a proportional increase in area in the basin and therefore shows negative correlation. The multiple regression demonstrates the degree of correspondence between variables in correlation coefficient. The net sown area shows high multiple correlation value of 0.633 cotton reveals high degree of association with net sown area of value of 0.864. Local jowar, in multiple regression points for limitations imposed by cotton and wheat variables indicating multiple value of 0.9236. Factor Analysis searches for common variances among variables. Four factors have been identified, with accounting for 85.3 per cent of the total variances. These four factors are labelled. (*1*) Cotton Factor, (*2*) Density Factor, (*3*) Pulses and oilseeds Factor, and (*4*) Wheat and tur Factor. The scores on Factors I, II, III and IV present spatial distribution. The mapping of spatial distribution of scores, indicates high positive loadings on cotton and net sown area in central part. The second factor exhibits positive loadings on irrigation, population density and agricultural density. Factor III reveals vivid picture of spatial distribution of scores on positive loadings on other pulses and sesamum, while the fourth factor shows positive loadings on wheat and tur in various patches within the Yashoda basin. The significant factors have been used for demarcating agricultural regions as it accounts for 89.7 per cent of the total explained variances by first two factors, *namely*, cotton and density. The scores on these two factors grouped 155 villages into three categories which are labelled as backward, developing and developed agricultural region. The class of backward regions is really a region of backward having less influence of cotton and density factor in fiftytwo villages (33.54 per cent to total villages dominating the western and northwest parts of the basin. The fiftyseven villages have been grouped as developing agricultural region as these villages are in the process of change and development. Developing agricultural region has relatively greater influence of cotton and density factor than that of backward agricultural region and is confined in the central part while developed agricultural region shows comparatively more influence of cotton and density factors. This region is highly dominated by cotton and density on medium to deep black soils in the central and southern parts including fortysix villages (29.67 per cent to total villages). Backward and developing agricultural regions can be brought to developed agricultural region by extending more hectarage under cultivation, use of improved seeds, application of fertilizers and pesticides, improved agricultural implements, and by increasing awareness among farmers about the method and techniques of agricultural practice in order to get high yield per hectare. This high yield of crops depends on farmers' decision making and therefore, efforts should be in this respect and directions.

7. The sample study of chosen villages uncovers indepth and comprehensive analysis of agricultural landuse pattern. The selected villages in present study exhibit some common characteristics. It is observed that agriculture is the basic source of livelihood and therefore more than 80 per cent working force to total workers are engaged in agriculture activity. Kharif is most important season in all sample villages when nearly 70 per cent to total enable land is brought under cultivation. Cotton and hybrid jowar are dominant in cropping pattern and stand first and second in sample villages. The agricultural landuse of individual villages undertaken for study exhibit variations due to different soil types and slight variations in rainfall amount, besides, decision making of farmers. It is found that Sarur and Rohankheda are relatively advanced in agricultural landuse than that of other sample villages because of their location in lower Yashoda basin, where medium to deep black soils are predominant and hence fertile producing high yield per hectare. The villages *namely*, Amboda and Nagazari have shown little development due to extensive coverage by coarse shallow soils. It is observed that fertile soils are utilized for cultivating cotton and hybrid jowar while, mug, mot, local jowar, groundnut are raised on inferior soil. The field survey, interviews and soil analysis unfold the association between soil types and crops for six sample villages as shown in the table.

Further it is observed that sample villages have brought maximum land under cultivation and therefore has less scope for explanation of arable land. The only possibilities of raising production is the intensification of landuse by sowing improved varieties of seeds, expansion of well irrigation, application of chemical fertilizers, in addition to awareness programme for propagation of modernisation.

Association of Soil Types and Crops

Sr. No.	*Name of villages*	*Soil types*	*Crops*
1.	Amboda	(1) Sanoy clay (2) Clay loam	Local jowar, mug, cotton and hybrid jowar
2.	Nagazari	(1) Silty loam	Cotton
		(2) Clay loam	Wheat
		(3) Clay	Hybrid jowar
3.	Kachangaon	(1) Sandy loam	Mot and mug
		(2) Clay loam	Hybrid jowar
		(3) Clay	Cotton
4.	Kurzadi	(1) Silty loam	Local jowar
		(2) Clay loam	Hybrid jowar
		(3) Clay	Cotton
5.	Sarur	(1) Sandy clay loam	Mot
		(2) Clay loam	Cotton and hybrid jowar
		(3) Clay	Hybrid jowar
6.	Rohankheda	(1) Silty loam	Cotton and hybrid jowar
		(2) Clay	Local jowar

CONCLUDING REMARK

The past investigations conducted in Maharashtra do not throw sufficient light on micro level analysis. Such studies have been attempted at taluka level only. The present work is an humble endeavour in analysis of agricultural landuse pattern in the Yashoda basin at the village level. This study may attract the attention of planners, agriculturists in particular and laymen in general. This exercise spotlights the past and present characteristics of agricultural landuse pattern of the basin. Such investigations would be of immense importance as guidelines for proper and efficient agricultural landuse for increasing production. Lastly it is felt that similar studies should be undertaken at the village level considering village as representative for further study for the Vidarbha region and other regions of Maharashtra, for in depth understanding of the agricultural landuse and its problems, in the near future.

A Selected Bibliography

Ahamad, A. and Siddiqui, M.F. 1967, "Crop Association Pattern in the Luni Basin," *Geographer*, Aligarh, 14/2.

Alexander, J.W. 1965, "Economic Geography", Prentice Hall, New Jersey.

Ali, Mohammad, 1978, "Studies in Agricultural Geography," Rajesh Publication, New Delhi.

Ali Mohammad, 1979, "Dynamics of Agricultural Development in India" (Ed.), Concept Publication, New Delhi.

Amani, K.Z. and Noor Mohammad, 1970, "Crop Combination in Trans-Trans-Gharghara Plaid", *Geographical Review of India*, Calcutta, 32/1.

Anderson, J.R., 1970, "A Geography of Agriculture Wm." (Brown Company, Dubuane, Iowa (U.S.A.).

Athawale, A.G., 1966, "Some New Methods of Crop-combination," *Geographical Review of India*, Calcutta, 28/3.

Ayyar, N.P., 1969, "Crop Regions of Madhya Pradesh: A study in methodology," *Geographical Review of India*, Calcutta, 31/1.

Baker, O.E., 1926, "Agricultural Regions of North America" *Economic Geography*, Vol. 2/1.

Banerjee, B., 1969, "Essays on agricultural Geography" Nabasakti Press (Ed.), Calcutta.

Bansil, P.C., 1977, "Agricultural Problems of India", Vikas Publication, New Delhi.

Basu, S., 1988, "Landuse in Malkauria village - A case study in Banker District", *Geographical Review of India*, Calcutta, 50/5.

Bhat, L.S.,1975, "Micro-level Planning", K.B. Publication, New Delhi.

Bhatia, S.S., 1960, "An Index of Crop-diversification." *Professional Geographer*, New York, 12/1.

Bhatia, S.S., 1965, "Pattern of Crop-concentration and diversification in India". *Economic Geography*, Worcester, 41/1.

Bhatia, C.R., 1981, "Changing Landuse and Cropping Pattern in Bihar", Perspective in Agricultural Geography, Concept Publication, New Delhi.

Chauhan, D.S., 1966, "Studies in Utilization of Land" Agra Shivalal Agrawal Company, Agra.

Chourudule, P.B., 1975, "Rural Landuse: A study in methods", *The Deccan Geographer*, Secunderabad, 13/1 & 2.

Coppock, J.T., 1968, "The Geography of Agriculture", *The Journal of Agricultural Economics*, England, 19/2.

Das, M.M., 1982, "Agricultural Landuse and Cropping Pattern," *Geographical Review of India*, Calcutta, 44/4.

Das, M.M., 1984, "Crop-combination Regions of Assam". A Quantitative Analysis", *National Geographer of India*, Varanasi, 30/3.

Datye, V.S., 1984, "Spatial Analysis of Agricultural Landuse in Poona District", Unpublished Ph.d. Thesis, Univ. of Poona, Pune.

David, W., 1966, "Theoretical Concept and Analysis of Agricultural Landuse in Geography", Review Article, *Geographer*, Aligarh, 50/2.

Deshmukh, V.M., 1975, "Lonkheda: A Study in Changing Landuse," *The Deccan Geographer*, Secunderabad, 13/1 & 2.

Doi, K., 1957, "The Industrial Structure of Japanese", Pretectures, Proceedings of the IGU, Regional Conference in Jannan.

Dikshit, K.R., 1973, "Agricultural Regions of Maharashtra," *Geographical Review of India*, Calcutta, 35/4.

Fireman, D., 1952, "Land Utilization Survey in Sehore", *Geographer*, Aligarh, 5/2.

Found, W.C., 1971, "Theoretical Approach to Rural Landuse Patterns", St. Martin, New York.

Fryer, D.W., 1965, "World Economic Development", McGraw Hill Book Company, New York.

Ganguli, B., 1953, "Landuse Survey and Agricultural Planning, in Uttar Pradesh," *Geographical Review of India*, Calcutta, 28/2.

Gondhalekar, Y.S., 1976, "Spatial Analysis of Yield of Major Crop of Maharashtra in relation to climatic condition", Unpublished Ph.D. Thesis, Univ. of Poona, Pune.

Government of India, 1974, Gazetteer of Wardha District.

Government of Maharashtra, 1971, Wardha District Census Handbook, Wardha.

Government of Maharashtra, 1981, Wardha District Census Handbook, Wardha.

Government of Maharashtra, 1989, Socio Economic Abstract for 1971 to 1989, Wardha District.

Geregor, H.F., 1970, "Geography of Agriculture Themes in Research", Prentice-Hall, New Jersey.

Grigg, D.B., 1969, The Agricultural System of the World. *Economic Geography*, 49/2.

Gupta, A.L. and Hiran, S.L. 1973, Agricultural Regions of Rajasthan, *The Deccan Geographer*, Secunderabad, 11/2.

Gupta, S.L., 1971, "Cropping Pattern of Haryana" *Geographical Review of India*, Calcutta, 33/1.

Hussain, M., 1972, "Crop Combination Regions of Uttar Pradesh: A Study in Methodology," *Geographical Review of India*, Calcutta, 38/1.

Husain, M., 1960, "Agricultural Adoges of Upper-Ganga Yamuna Doab", *The Geographer*, Aligarh, 16.2.

Hussain, M., 1979, "Agricultural Geography," Inter-India Publication, New Delhi.

Jana, M. and Baghi, K. (1984) 1984, "The Crop Combination and Spatial Pattern of Land Utilization in Lower Silabati Basin," *Geographical Review of India*, 36/4.

Joshi, Y.G., 1972, "Agricultural Geography of the Narmada Basin" (Hindi), Madhya Pradesh Academi, Bhopal.

Kadam, N.M., 1991, Ahmednagar District - A Study in Regional Planning, Unpublished Ph.D. Thesis, Univ. of Poona, Pune.

Karmarkar, P.R., 1981, "Agricultural Landuse of the Western Districts of Upland Maharashtra," Unpublished Ph.D. Thesis, University of Poona, Pune.

Khatu, K.K., 1975, Baroda Slums: A case study in misuse of Land, *The Deccan Geographer*, Secunderabad, 13/2.

Kumbhare A.R., 1978, Agricultural Landuse in Upper Godavari Basin. The Unpublished Ph.D. Thesis, University of Poona, Pune.

Kuniyal, J.C., 1987, "Crop Concentration and Diversification in Nainital District, Uttar Pradesh," Himalaya, *National Geographical Journal of India*, New Delhi, 33/2.

Mohammad, N. and Amani, K.Z., 1970, "Crop Combination in Trans-Ghagara Plain," *Geographical Review of India*, Calcutta, 32/1.

Mathur, R.N. and Prasad, R., 1981, Impact of Consolidation of Agricultural Landuse Pattern," *Perspective in Agricultural Geography* (Ed.), Noor Mohammed, Concept Publication, New Delhi.

Mustafa, F.R. and Development in Maharashtra, *Transactions*, Pune, 6/2.

Morgan, W.B. and Munton, R.J., 1972, "Agricultural Geography," Methuen and Company Ltd., London.

Mukherjee, A.B., 1963, "Landuse Pattern in Hewar Village, Rajasthan", *Geographical Review of India*, Calcutta, 30/2.

Mukherjee, S.N., 1967, "Agricultural Planning in Howrah," Geographical Review in Mewar Village, Rajasthan", *Geographical Review of India*, Calcutta, 29/2.

Noor Mohammad, 1970, "Crop-combination in Trans-Ghagara Plain", *Geographical Review of India*, Calcutta, 32/1.

Noor Mohammad, 1973, Misraulia: A study in Agricultural Landuse (1951-70). *Geographical Review of India*, Calcutta, 35/3.

Noor Mohammad, 1981, "Patterns of Agricultural Landuse in Geographer-Rapti Doab" *Perspective in Agricultural Geography*, (Ed.), Noor Mohammad, Concept Publication, New Delhi.

Noor Mohammad, 1981, *Perspective in Agricultural Geography* (Vol. 1 to 5) (Ed.) Concept Publication, New Delhi.

Profali, R.M., 1975, "The Study of Land Utilization in a part of Bor Command Area." *The Deccan Geographer*, Secunderabad, 13/1 & 2.

Rafullah, S.N. 1967, "Region Regionalism and Regionalization," *The Geographer*, Aligarh, 14/1.

Rao, S.K., 1971, "Inter-Regional Variations in Agricultural Growth (1952-53 to 1964-65)." *Economic and Political Weekly*, Bombay, 6/27.

Reddy, P.V., 1983, "Pattern of Croplanduse in Chandragiri Taluka, Chittor District," Andhra Pradesh, *National Geographical Journal of India*, New Delhi, 29/1 & 3.

Shafi, M., 1965, "Pattern of Croplanduse in Ganga-Yamuna Doab." *The Deccan Geographer*, Secunderabad, 13/1 & 2.

Sharma, B.M.P., 1975, "The Cropping Pattern in the Upper Tons Basin," *The Deccan Geographer*, Secunderabad, 13/1 & 2.

Sharma, J.K., 1987, "Agricultural Land Utilization, Planning and Development of Dharwar Taluka, Karnataka," *National Geographical Journal of India*, New Delhi, 33/2.

Sharma, P.R., 1072, "Crop Cultivation, Intensity, their ranking and Crop Association Regions, in Chattisgarh Region, A Geographical Analysis," *National Geographical Journal of India*, Varanasi, 18/2.

Sharma, T.C., 1972, "Pattern of Croplanduse in Uttar Pradesh." *The Deccan Geographer*, Secunderabad 10/1.

Sharma, B.L., 1991, "Pattern of Croplanduse in Uttar Pradesh." *The Deccan Geographer*, Secunderabad 10/1.

Sharma, S.C., 1971, "Cropping Pattern and Crop-combination Regions in Etweal Middle Ganga Yammuna Doab," *The Deccan Geographer*, Secunderabad, 9/1.

Shastri, P., 1981, "Changes in Landuse and Cropping Pattern in the Cotton Belt of Vidharbha," *Perspective in Agricultural Geography* (Ed.), Noor Mohammad, Concept Publication, New Delhi.

Shenoi, P.V., 1975, "Agricultural Development in India," Vikas Publishing House, New Delhi.

Shinde, S.D., 1980, "Agriculture in an underdevelopment region — a Geographical Survey," Himalaya Publication, Bombay.

Shinde, S.D., 1973, "The Panchaganga Basin: An Approach of Some Aspects of its Agricultural Geography". *Geographical Review of India*, Calcutta, 35/3.

Siddique, M.F., 1973, "Agricultural Landuse in Black soil region of Bundelkhad." *Geographical Review of India*, Calcutta, 35/4.

Siddique, M.F., 1964, "Combinational Analysis, A Review of Methodology," *The Geographer*, Aligarh, 14/2.

Singh, B.B., 1967, "Landuse Cropping Pattern and their Ranking." *The National Geographical Journal of India*, Varanasi, 13/2.

Singh, G.B., 1979, "Transformation of Agriculture," Vishal Publication, Kurukshetra, Haryana.

Singh, H., 1963, "Crop Diversification in Malwa Tract of Panjab." *The Indian Geographical Journal*, Madras, 38/3 & 4.

Singh, J. and Dhillon, S.S., 1984, "Agricultural Geography," Tata-McGraw-Hill, Publishing Company Limited, New Delhi.

Singh, J., 1976, "Agricultural Atlas of India," Vishal Publication, Kurukshetra, Haryana.

Singh, J., 1976, "An Agricultural Geography in Haryana," Vishal Publication, Kurukshetra, Haryana.

Singh, T., 1977, "Bihar: A Study of Crop-combination Regions." *Indian Journal of Regional Science*, Kharagpur, 9/2.

Singh, S., 1967, "Land utilization in Dunaivillage, Aligarh District," *The Deccan Geographer*, Secunderabad, 5/1 & 2.

Sohal, K.S. and Khaira, 1989, Temoral Changes in Crop-combination Regions in Sangror District,'' Panjab, *National Geographer*, Allahabad, 24/2.

Stamp, L.D., 1962, "The Land of Britain: Its Use and Misuse." Third Edition, London.

Symons, L., 1967, "Agricultural Geography", G. Bell & Sons, London.

Torrant, J.R., 1974, "Agriculture, Geography," John Wiley and Sons, New York.

Vaidya, B.C., 1990, "Cropping pattern in the Yashoda Basin." *The Deccan Geographer*, Secunderabad, 28/2 & 3.

Vaidya, B.C., 1885, "Impact of Physical Environment on Agricultural Landuse - A Case Study of Mula Basin," Poona District, Maharashtra, Unpublished M.Phil. Dissertation, Univ. of Poona, Pune.

Vaidya, B.C., 1992, "Geographical Analysis of Agricultural Landuse at Kachangaon in Yashoda Basin in Maharashtra". (Hindi). Maharashtra Bhugolshustra Sanshodhan Patrika, Pune 6/1.

Vaidya, B.C., 1989, "Pattern of Crop-diversification in the Yashoda Basin Maharashtra," (Marathi), Maharashtra Bhugol Shastra Sanshodhan Patrika, Pune, 3/1.

Vaidya, B.C., 1991, "Temporal Change in the Cropping Pattern of Cotton Growing Region of the Yashoda Basin, Maharashtra," *National Geographer*, Allahabad, 26/1.

Vink, A.P.A., 1975, "Landuse in Advance Agriculture." Springer-Verlag.

Wadia, D.N., 1953, "Geology or India," Mac-Millan, London.

Weaver, J.C., 1954, "Changing Pattern of Croplanduse in the Middlewest." *Economic Geographer*, Worcester, 30/2.

Weaver, J.C., 1954, "Crop-combination Regions in the Middleast." *Geographical Review*, XL/V, New York.

Whittlessey, D., 1936, "Major Agricultural Regions of the Earth," *Annals of American Geographers*, 25/1.

INDEX